# Iwasawa Theory of Elliptic Curves with Complex Multiplication

**PERSPECTIVES IN MATHEMATICS, Vol. 3**
J. Coates and S. Helgason, editors

# Iwasawa Theory of Elliptic Curves with Complex Multiplication

## *p*-adic *L* Functions

Ehud de Shalit

*Mathematical Sciences Research Institute*
*Berkeley, California*

ACADEMIC PRESS, INC.
*Harcourt Brace Jovanovich, Publishers*

Boston Orlando San Diego
New York Austin London Sydney
Tokyo Toronto

ACADEMIC PRESS, INC.
Orlando, Florida 32887

*United Kingdom Edition published by*
ACADEMIC PRESS, INC. (LONDON) LTD.
24-28 Oval Road, London, NW1 7DX

Library of Congress Cataloging-in-Publication Data
De Shalit, Ehud.
Iwasawa theory of elliptic curves with complex multiplication.

(Perspectives in mathematics; vol. 3)
Bibliography: p.
1. L-functions. 2. Curves, Elliptic. 3. Class field theory. I. Title. II. Series.
QA247.D36 1987 512'.74 87-1435
ISBN 0-12-210255-X (alk. paper)

87 88 89 90 9 8 7 6 5 4 3 2 1
Printed in the United States of America

**TO THE MEMORY OF MY FATHER**

TABLE OF CONTENTS

## ACKNOWLEDGEMENTS

This book, based in part on the author's Ph.D. thesis, summarizes the work of many people, and benefited from diverse sources. I tried to give appropriate credits and references in the text, but there are undoubtedly omissions of which I remain unaware.

Like everybody else in the field, I was heavily influenced by the beautiful ideas of J. Coates and A. Wiles, and by the analogy with Iwasawa's work on cyclotomic fields. In addition, Iwasawa's approach to local class field theory, and the work of R. Coleman, formed the basis of chapter I. Chapter II grew out of, and is a natural continuation to the work of R. Yager. Part of it was planned as a joint paper with him, but due to geographical difficulties, this plan never materialized. To all these people I would like to express my deepest gratitude.

Special thanks go to K. Iwasawa for guiding me in my first steps, to Andrew Wiles for his encouragement and friendship, and to Robert Coleman who read parts of the manuscript and made valuable comments.

I would like to thank all the other teachers and friends from whom I learned the subject, in particular E. Friedman, R. Greenberg, B. Gross, B. Mazur and K. Rubin.

I am also grateful to Harvard University, the Mathematical Sciences Research Institute, and the National Science Foundation for their support during the preparation of this book, and to Faye Yeager for the excellent typing.

Ehud de Shalit

Berkeley, December 8, 1986

# INTRODUCTION

$p$-adic $L$ functions are analytical functions of $p$-adic characters that, one way or another, interpolate special values of classical (complex) $L$ functions. The first such examples were the $p$-adic $L$ functions of Kubota and Leopoldt [K-Le], interpolating Dirichlet $L$ series. Manin and Vishik [M-V] and Katz [K1] constructed $p$-adic $L$ functions which interpolate special values of Hecke $L$ series associated with a quadratic imaginary field $K$, in which $p$ splits. (To fix notation write $p = \mathfrak{p}\bar{\mathfrak{p}}$). Their work gave $p$-adic interpolation of the Hasse-Weil zeta function of certain elliptic curves with complex multiplication and good ordinary reduction at $\mathfrak{p}$ (those whose division points generate abelian extensions of $K$). The $p$-adic $L$ function of Manin-Vishik and Katz is the first object studied in this work.

Our point of view is nevertheless different, and goes back to the two fundamental papers [C-W1] and [C-W2] by Coates and Wiles. The program, pursued by various authors since (see the introduction to chapter II), and which is brought here to its fullest generality (so we hope), may be summarized in two main steps.

Fix an abelian extension $F_1$ of $K$, and let $K_\infty$ be the unique $\mathbf{Z}_p$ extension of $K$ unramified outside $\mathfrak{p}$ (one of the two factors of $p$ in $K$). If we assume that $F_1$ is the ray class field modulo $\mathfrak{f}\mathfrak{p}$, where $\mathfrak{f}$ is an integral ideal relatively prime to $\mathfrak{p}$, we do not lose any generality, and some notation is simplified. We therefore make this assumption. The $p$-adic $L$ function, then, is essentially a $p$-adic integral measure on $\mathcal{G} = Gal(F_1K_\infty/K)$.

Now in the first step we are given a norm-coherent sequence $\beta$ of semi-local units in the completion of the tower $F_\infty = F_1K_\infty$ at $\mathfrak{p}$. Out of each such sequence we construct a certain measure $\mu_\beta$ on $\mathcal{G}$. We describe this construction in chapter I. In the second step, carried out in chapter II, we introduce special global units, the elliptic units. They come in norm coherent sequences, so we can view them inside the local units. When the procedure from chapter I is applied to them we obtain the $p$-adic $L$ function.

Chapters I and II are carried out in full generality and are also attempted to be self contained. This results in long tedious computations. The reader who approaches the subject for the first time is advised to make two simplifying assumptions: that $K$ is of class number 1, and that the grossencharacters in question are unramified at $\mathfrak{p}$. These eliminate most of the technical difficulties, yet very little is lost conceptually. If still confused, one may restrict attention to grossencharacters of infinity type $(k, 0)$. This will only give the interpolation formula for the "one variable" $p$-adic $L$ function. We have actually treated this case separately in II.4, despite some repetition, to facilitate the reading.

Other results obtained in chapters I and II include a new proof of Wiles' explicit reciprocity law, a $p$-adic analogue of Kronecker's limit formula, and a functional equation for the $p$-adic $L$ function.

The immense interest in Katz' $p$-adic $L$ functions arises from their significance to class field theory (abelian extensions of $K$) and the arithmetic of elliptic curves with complex multiplication. In the last two chapters we give a sample of results in these two directions. Although largely self-contained, these chapters are not intended to be exhaustive, and several topics are omitted. The selection of material, and sometimes the method of proof, were influenced by our desire to show how the results of chapter II are put to use.

Chapter III is mainly concerned with the "main conjecture" in the style of cyclotomic Iwasawa theory. The fundamental idea is that the zeroes of the $p$-adic $L$ function ought to be those $p$-adic characters of $\mathcal{G}$ whose reciprocals appear in the representation of $\mathcal{G}$ on a certain free $\mathbf{Z}_p$-module of finite rank. More precisely, this module $\mathcal{X}$ is the Galois group of the maximal abelian $p$-extension of $F_\infty$ which is unramified outside $\mathfrak{p}$. See the introduction to chapter III for more details. We prove that the Iwasawa invariants of $\mathcal{X}$ and the Iwasawa invariants of the $p$-adic $L$ function are equal, but we do not go into the recent evidence for this conjecture discovered by K. Rubin, nor do we give Gillard's proof of the vanishing of the $\mu$-invariant.

While elliptic curves are deliberately kept behind the scene in chapter III, their arithmetic, and in particular the conjecture of Birch and Swinnerton-Dyer, is the main topic of chapter IV. First we show how Kummer theory and descent are used to relate the Galois group previously denoted by $\mathcal{X}$ to the Selmer group over $F_\infty$. Then we give a complete proof of two beautiful theorems of Coates-Wiles and of R. Greenberg. These theorems are generalized here to treat elliptic curves with CM by an arbitrary quadratic imaginary field, not necessarily of class number 1. The crucial hypothesis that must be kept is that the division points of the curve in question generate an abelian extension of $K$.

Of the topics not considered here, let us mention $p$-adic heights and $p$-adic sigma functions, the work of Perrin-Riou on the algebraic analogue of the conjecture of Birch and Swinnerton-Dyer [PR1], and her "Gross-Zagier-type" result [PR2]. As this book goes to press, K. Rubin has announced important new results concerning the conjecture of Birch and Swinnerton-Dyer. He kindly allowed me to report on them here, and we refer the reader to his forthcoming papers for details.

The author is well aware of the lack of numerical examples in chapters III and IV. These would illustrate the theory magnificently, but due to lack of skill in computing, I was unable to produce any new examples. There is much relevant numerical data in the paper of Bernardi, Goldstein and Stephens [B-G-S].

# CHAPTER I

# FORMAL GROUPS, LOCAL UNITS, AND MEASURES

Much of the first half of this book is devoted to the construction of $p$-adic $L$ functions associated with quadratic imaginary fields. This construction is "formal" and "local" in the beginning. Only at a later stage results from the theory of complex multiplication are incorporated. In chapter I we gather those results which *do not* deal with elliptic curves. Our tools are formal groups and $p$-adic measures. The key result is theorem 3.7, which describes the structure of a certain module of local units. This module plays a central role in the following three chapters. In section 4 we prove a version of the explicit reciprocity law in local class field theory, that will be needed in chapter IV.

## 1. Relative Lubin-Tate Groups

**1.1** Let $R$ be a commutative ring with identity. For our purpose a (commutative) *one dimensional formal group law* over $R$ is a power series $F \in R[[X,Y]]$, satisfying the following axioms.

(i) $F(X,Y) \equiv X + Y \ mod\ deg\ 2$
(ii) $F(X,0) = X = F(0,X)$
(iii) $F(X,F(Y,Z)) = F(F(X,Y),Z)$ (associativity)
(iv) $F(X,Y) = F(Y,X)$ (commutativity).

We use the notation $f \equiv g \ mod\ deg\ n$ to mean that $f - g$ involves only monomials of total degree not less than $n$. It can be shown ([Haz] 1.1.4) that there exists a unique power series $\iota(X) \in R[[X]]$ such that $F(X,\iota(X)) = 0$.

Let $A$ be an $R$-algebra and $\mathfrak{a}$ an ideal such that $A$ is complete and separated in its $\mathfrak{a}$-adic topology (i.e. $A = \varprojlim A/\mathfrak{a}^n$). Then if $f,g \in \mathfrak{a}$, $F(f,g)$ and

$\iota(f)$ converge to elements of $\mathfrak{a}$. We denote them by $f[+]g$ and $[-]f$ respectively, and observe that with $[+]$ as addition $\mathfrak{a}$ becomes an abelian group, written $F(\mathfrak{a})$ ("the $\mathfrak{a}$-valued points of $F$"), to distinguish from the ordinary addition on $\mathfrak{a}$. These remarks apply in particular to $A = R[[X]]$ and $\mathfrak{a} = (X)$, and to the case where $A = R$ is a complete local ring and $\mathfrak{a}$ is its maximal ideal. Almost everything we shall need about formal groups can be found in the book of Hazewinkel [Haz]. Henceforth we let "formal group" stand for "commutative one-dimensional formal group law", unless otherwise specified.

A *homomorphism* $f$ between two formal groups $F$ and $F'$ over $R$ is a power series without constant term such that $F'(f(X), f(Y)) = f(F(X,Y))$. The collection $Hom(F, F')$ of such homomorphisms forms a group with respect to the addition law of $F'$ : $(f+g)(X) = f(X)[+]'g(X)$, and $End\,(F)$ becomes a *ring* under *composition* as product.

Let $R$ be a domain of characteristic 0, and $f \in Hom(F, F')$. Then $f(X) = aX +$ (higher terms) and the map $f \mapsto a = f'(0)$ is an *injective* group homomorphism of $Hom(F, F')$ into $R$ ([Haz] 20.1). When $F = F'$ this is a ring homomorphism. We shall write $[a]_{F,F'}$ or $[a]_F$, or simply $[a]$ instead of $f$ in such a case. Over the field of fractions $K$ of $R$ all formal groups are isomorphic. Any isomorphism $\lambda : F \simeq \hat{\mathbf{G}}_a$ over $K$ ($\hat{\mathbf{G}}_a(X,Y) = X + Y$ is the *additive formal group*) is called a *logarithm* of $F$. If $\lambda$ is normalized so that $\lambda'(0) = 1$, then $\lambda'(X) \in R[[X]]^x$ has coefficients in $R$ ([Haz] 5.8). All these statements are blatantly false (or void) in non-zero characteristic.

Let $F$ be a formal group over a field of characteristic $p > 0$. Then $[p]_F(X) = X[+]\ldots[+]X$ ($p$ times) is a power series in $X^q$ with $q = p^h$ for some $h \geq 1$. The largest possible $h$ is called the *height* of $F$ ([Haz] 18.3). If $[p]_F = 0$ $F$ is of infinite height.

Finally, we shall need the concept of a *translation-invariant derivation* on $F$. This is a continuous derivation $D$ of $R[[X]]$ (over $R$) satisfying $D(f(X[+]Y)) = Df(X[+]Y)$, where $Y$ is treated here as a constant for $D$ (i.e. $D$ is extended to

$R[[X,Y]]$ via $DY = 0$). If $R$ is a domain of characteristic 0, then $D = \frac{c}{\lambda'(X)} \frac{d}{dX}$ where $c \in R$ and $\lambda$ is the logarithm of $F$, normalized to $\lambda'(0) = 1$.

The *multiplicative formal group* $\hat{\mathbf{G}}_m$ is given by $\hat{\mathbf{G}}_m(X,Y) = X + Y + XY = (1 + X)(1 + Y) - 1$.

**1.2** Let $k$ be a finite extension of $\mathbf{Q}_p$, the field of $p$-adic numbers. Let $\mathcal{O}$ and $\wp$ be its valuation ring and maximal ideal. Let the residue field $\mathcal{O}/\wp$ have $q$ elements. Lubin and Tate introduced an extremely useful class of formal groups defined over $\mathcal{O}$ [L-T]. Their handiness stems from the fact that they each possess a special endomorphism which "lifts" the Frobenius substitution $X \mapsto X^q$ in characteristic $p$. Here we generalize a little (see [dS1]), and as usual in this theory, focus first on the lifting of Frobenius, and web the formal group around it.

Let $d$ be a positive integer and $k'$ the unique unramified extension of $k$ of degree $d$. Let $k^{ur}$ be the maximal unramified extension of $k$, and $K$ its completion. The Frobenius automorphism (relative to $k$) $\varphi$ generates $Gal(k^{ur}/k)$ topologically, and extends by continuity to $K$. It is characterized by $\varphi(x) \equiv x^q \; mod \; \wp^{ur}$ for all $x \in \mathcal{O}^{ur}$. We let $\mathcal{O}', \wp', \varphi'$ denote the corresponding objects for $k'$, so that $\varphi' = \varphi^d$. Finally let $\nu : K^x \to \mathbf{Z}$ be the normalized valuation (normalized in the sense that $\nu(K^x) = \mathbf{Z}$).

Fix $\xi \in k^x$, $\nu(\xi) = d$, and consider

$$\mathcal{F}_\xi = \{f \in \mathcal{O}'[[X]] \mid f \equiv \pi' X \; mod \; deg \; 2, \; N_{k'/k}(\pi') = \xi \text{ and } f \equiv X^q \; mod \; \wp'\}.$$

Any $f$ in $\mathcal{F}_\xi$ is going to play the role of an endomorphism lifting Frobenius. Its differential is $f'(0) = \pi'$, and its reduction is $X^q$.

**1.3 Theorem.** *For every $f \in \mathcal{F}_\xi$ there exists a unique one-dimensional commutative formal group law $F_f$ defined over $\mathcal{O}'$ satisfying $F_f^\varphi \circ f = f \circ F_f$.*

In other words, $f \in Hom(F_f, F_f^\varphi)$. Here, and elsewhere, the superscript $\varphi$ means that we apply $\varphi$ to the coefficients of the power series. Note that $F_f^\varphi \in \mathcal{F}_\xi$ too, and $F_f^\varphi = F_{\varphi(f)}$ (apply $\varphi$ to the equation defining $F_f$). When $d = 1$ we are in the situation studied by Lubin and Tate. When $d \geq 1$ we call $F_f$ a *relative Lubin Tate* group (relative to the extension $k'/k$). For the proof we need a lemma.

**1.4 Lemma.** *Let $f, g \in \mathcal{F}_\xi$ and let $F_1(X_1, \ldots, X_n)$ be a linear form in $\mathcal{O}'[X_1, \ldots, X_n]$. Suppose $f \circ F_1 \equiv F_1^\varphi \circ (g, \ldots, g)$ mod deg 2. Then there exists a unique $F \in \mathcal{O}'[[X_1, \ldots, X_n]]$ satisfying (i) $F \equiv F_1$ mod deg 2, (ii) $f \circ F = F^\varphi \circ (g, \ldots, g)$.*

PROOF: (Compare [Se] p. 149). Let $f = \pi_1 X + \ldots, g = \pi_2 X + \ldots$. Set $F^{(1)} = F_1$ and define successive approximations $F^{(m)}$ satisfying (ii) *mod deg* $m+1$ through $(m \geq 2)$ $F^{(m)} = F^{(m-1)} + F_m$ where $F_m$ is homogeneous of degree $m$. For this we need

$$f \circ \left(F^{(m-1)} + F_m\right) \equiv \left(F^{(m-1)} + F_m\right)^\varphi \circ g \textit{ mod deg } m + 1$$

or

$$f \circ F^{(m-1)} + \pi_1 F_m \equiv F^{(m-1)\varphi} \circ g + \pi_2^m F_m^\varphi.$$

Let $t$ be the homogeneous part of degree $m$ of $F^{(m-1)\varphi} \circ g - f \circ F^{(m-1)}$. Since $F^{(m-1)\varphi} \circ g \equiv F^{(m-1)\varphi}(X_1^q, \ldots, X_n^q) \equiv (F^{(m-1)})^q \equiv f \circ F^{(m-1)}$ *mod* $\wp'$, $t \equiv 0$ *mod* $\wp'$. We have to find $F_m$ satisfying

$$F_m - \pi_1^{-1}\pi_2^m F_m^\varphi = \pi_1^{-1} t.$$

This is possible because $m \geq 2$ and $\mathcal{O}'$ is complete (proceed by induction *mod* $\wp'^r$). Setting $F = \sum_{m=1}^{\infty} F_m$ concludes the proof.

PROOF OF THEOREM 1.3: In the lemma, let $f = g$ and $F_1 = X_1 + X_2$. We have to show that $F_f$ = the resulting $F$, is a formal group law. This is done by repeated application of lemma 1.4 and is left as an exercise (or look it up in [Se] p. 150).

Let $\tilde{F}$ be the reduction of $F_f$, i.e. the formal group over $\mathcal{O}'/\wp'$ obtained by "reading $F_f$ modulo $\wp'$". It is easily verified that $\tilde{F}$ is of height $[k : \mathbf{Q}_p]$. By abuse of language we refer to it as the *height* of $F_f$ too.

**1.5 Proposition.** *Let* $f = \pi_1 X + \cdots$, $g = \pi_2 X + \cdots$ *be in* $\mathcal{F}_\xi$. *Let* $a \in \mathcal{O}'$ *satisfy* $a^{\varphi-1} = \pi_2/\pi_1$ *(such an a exists by Hilbert's theorem 90). Then there exists a unique power series* $[a]_{f,g} \in \mathcal{O}'[[X]]$ *for which*

*(i)* $[a]_{f,g} \equiv aX$ *mod deg* 2,

*(ii)* $[a]^{\varphi}_{f,g} \circ f = g \circ [a]_{f,g}$.

*Also* $[a]_{f,g} \in Hom(F_f, F_g)$. *If* $h(X) = \pi_3 X + \cdots$ *is another element of* $\mathcal{F}_\xi$ *and* $b^{\varphi-1} = \pi_3/\pi_2$, *then* $[ab]_{f,h} = [b]_{g,h} \circ [a]_{f,g}$. *Moreover, the map* $a \mapsto [a]_{f,g}$ *is a group isomorphism from* $\{a \in \mathcal{O}' \mid a^{\varphi-1} = \pi_2/\pi_1\}$ *onto* $Hom(F_f, F_g)$, *and if* $f = g$ *it is a ring isomorphism* $\mathcal{O} \simeq End\,(F_f)$.

**Corollary.** *If* $f, g \in \mathcal{F}_\xi$, $F_f$ *and* $F_g$ *are isomorphic over* $\mathcal{O}'$.

PROOF: In lemma 1.4 let $F_1 = aX$, so that $g \circ F_1 \equiv F_1^{\varphi} \circ f$ *mod deg* 2. Then call the resulting $F$ $[a]_{f,g}$. Transitivity $[ab]_{f,h} = [b]_{g,h} \circ [a]_{f,g}$ is clear from the uniqueness. To show $F_g \circ [a] = [a] \circ F_f$ it is enough to check (in $k'[[X]]$, where the inverse is with respect to composition) that $g \circ [a] \circ F_f \circ [a]^{-1} = [a]^{\varphi} \circ F_f^{\varphi} \circ ([a]^{\varphi})^{-1} \circ g$, which follows from (ii) and the defining property of $F_f$. The only non-obvious point is that every homomorphism $\lambda : F_f \to F_g$ is of this form. But let $a = \lambda'(0)$. By the general remarks in 1.1 concerning the injectivity of the map $\lambda \mapsto \lambda'(0)$ in characteristic 0, it is enough to show that $a^{\varphi-1} = \pi_2/\pi_1$, which is obvious. Notice that even if we allow homomorphisms defined over the integers of an extension field of $k'$, we do not get anything new.

**1.6** It follows from the last section that the various $F_f$, for $f \in \mathcal{F}_\xi$, can be viewed as different *models* of the same formal group. Over $\mathcal{O}_K$ (recall $K = \hat{k}^{ur}$) one need not even distinguish between the various $\xi$. In other language, $\mathcal{O}_K$ is strictly Henselian so formal groups over it are canonically classified by their reduction (over $\overline{\mathbf{F}}_p$), which is determined up to isomorphism by its height. In down to earth terms, we have the following proposition.

**Proposition.** *Suppose* $\nu(\xi) = \nu(\xi') = d$, *set* $v = \xi'/\xi$, *and let u be a unit of* $k'$ *such that* $N_{k'/k}(u) = v$ *(the norm map on units is surjective in an unramified* *extension). Pick* $\Omega \in \mathcal{O}_K^x$ *such that* $\Omega^{\varphi-1} = u$, *and* $f \in \mathcal{F}_\xi$. *Then there exists a unique power series* $\theta(X) \in \mathcal{O}_K[[X]]$ *satisfying*

*(i)* $\varphi^d(\theta) = \theta \circ [v]_f$,

*(ii)* $\theta(X) = \Omega X +$ *(higher terms)*.

*Put* $f' = \theta^\varphi \circ f \circ \theta^{-1}$. *Then* $f' \in \mathcal{F}_{\xi'}$ *and* $\theta$ *is an isomorphism of* $F_f$ *onto* $F_{f'}$ *over* $\mathcal{O}_K$.

The proof is left to the reader. Note that $\Omega$ exists because $\mathcal{O}_K$ is complete and its residue field is algebraically closed.

**1.7** For $i \geq 0$ and $f \in \mathcal{F}_\xi$, let $f^{(i)} = \varphi^{i-1}(f) \circ \cdots \circ \varphi(f) \circ f$. Then $f^{(i)} \in Hom(F_f, F_{\varphi^i(f)})$, and if $\nu(\xi) = d$, $f^{(d)} = [\xi]_f \in End\,(F_f)$ (we write $[a]_f$ for $[a]_{f,f}$). Note also that $\varphi^j(f^{(i)}) \circ f^{(j)} = f^{(i+j)}$.

Let $\mathbf{C}_p$ denote the completion of an algebraic closure of $\mathbf{Q}_p$, $M$ its valuation ideal, and $M_f = F_f(M)$, the $M$-valued points of $F_f$.

DEFINITION: Let $\pi$ be any prime element in $\mathcal{O}$, $n \geq 0$. Define

$$\begin{aligned} W_f^n &= \{\omega \in M_f \mid [a]_f(\omega) = 0 \quad \text{for all } a \in \wp^n\} \\ &= \{\omega \in M_f \mid [\pi^n]_f(\omega) = 0\} \\ &= Ker(f^{(n)} : M_f \to M_{\varphi^n(f)}). \end{aligned}$$

These are the *division points of level n* in $F_f$. The equality of the first two expressions is obvious. They are also equal to the third one because, quite generally, if $a \in \mathcal{O}'$, $b \in \mathcal{O}$ and $\nu(a) = \nu(b)$, $[a]_{f,g} \in Hom(F_f, F_g)$, then $[a]_{f,g} = [ab^{-1}]_{f,g} \circ [b]_f$ and $[ab^{-1}]_{f,g}$ is an isomorphism.

**Proposition.** *(i)* $W_f^n$ *is a finite sub-*$\mathcal{O}$*-module of* $M_f$ *of* $q^n$ *elements.* $W_f^n \subset W_f^{n+1}$.

*(ii) If* $\omega \in \tilde{W}_f^n = W_f^n \backslash W_f^{n-1}$, $a \mapsto [a]_f(\omega)$ *gives an isomorphism* $\mathcal{O}/\wp^n \simeq W_f^n$.

*(iii)* $W_f = \cup W_f^n \cong k/\mathcal{O}$ *(noncanonically) and is the set of all* $\mathcal{O}-torsion$ *in* $M_f$.

The proof is the same as in [Se], §3.6, prop. 6, and we omit it.

**1.8 Proposition.** *The field $k'(W_f^n) = k_\xi^n$ does not depend on which $f$ we choose, as long as $f \in \mathcal{F}_\xi$. It is a totally ramified extension of $k' = k_\xi^0$ of degree $(q-1)q^{n-1}$ $(n \geq 1)$, which is abelian over $k$. Any $\omega \in \tilde{W}_f^n$ generates $k_\xi^n$ over $k'$ and is in fact a prime element in $k_\xi^n$. There is a canonical isomorphism $(\mathcal{O}/\wp^n)^x \simeq Gal(k_\xi^n/k')$ given by $u \mapsto \sigma_u$, $\sigma_u(\omega) = [u^{-1}]_f(\omega)$ $(\omega \in W_f^n)$, and this isomorphism is again independent of $f$.*

PROOF: See [Se] §3.6 and 3.7. The only point that deserves special attention in the "relative" situation is that $k_\xi^n$ is actually abelian over $k$, and not only over $k'$. First, since it is independent of $f$, it is clearly Galois over $k$. Let $\tau$ be an extension of $\varphi$ to $k_\xi^n$ and $\sigma \in Gal(k_\xi^n/k')$, say $\sigma = \sigma_u$. Then $\tau\sigma_u(\omega) = \tau([u^{-1}]_f(\omega)) = [u^{-1}]_{\varphi(f)}(\tau\omega) = \sigma_u(\tau\omega)$ since the isomorphism $u \mapsto \sigma_u$ is independent of $f$, which proves that $\sigma$ and $\tau$ commute.

See §4 for more about these fields and local class field theory. The map $u \mapsto \sigma_u$ is the well known local Artin symbol. A fact to bear in mind is that $k_\xi^n$ is class field to the subgroup $<\xi> \cdot (1 + \wp^n)$ of $k^x$ $(<\xi> \cdot \mathcal{O}^x$ if $n = 0)$. If $k''$ is an unramified extension of $k'$ of degree $e$, $k''k_\xi^n = k_{\xi'}^n$, for $\xi' = \xi^e$. In particular, for any $n$, $\xi_1$ and $\xi_2$, there exists $k''$ for which $k''k_{\xi_1}^n = k''k_{\xi_2}^n$. See also the forthcoming book of Iwasawa [Iw 3].

## 2. COLEMAN'S POWER SERIES

Robert Coleman introduced in [Col1] a method of dealing with norm-coherent sequences in the tower of local fields $k_\xi^n$. The Coleman power series is a simple but ingenious device which associates with any such sequence a power series over $\mathcal{O}'$.

**2.1** THE NORM OPERATOR: Notation as in §1 let $R = \mathcal{O}'[[X]]$, $\xi \in k^x$, $\nu(\xi) = d$, and $f \in \mathcal{F}_\xi$.

**Proposition.** *There exists a unique multiplicative operator $\mathcal{N} : R \to R$ ($\mathcal{N} =$*

$\mathcal{N}_f$ *when we want to emphasize the dependence on* $f$), *such that*

$$\text{(1)} \qquad \mathcal{N}h \circ f = \prod_{\omega \in W_f^1} h(X[+]\omega) \qquad \forall h \in R.$$

*(addition is on the formal group* $F_f$*, of course). It enjoys the additional properties:*

*(i)* $\mathcal{N}h \equiv h^{\varphi} \bmod \wp'$,

*(ii)* $\mathcal{N}_{\varphi(f)} = \varphi \circ \mathcal{N}_f \circ \varphi^{-1}$,

*(iii) Let* $\mathcal{N}_f^{(i)} = \mathcal{N}_{\varphi^{i-1}(f)} \circ \cdots \circ \mathcal{N}_{\varphi(f)} \circ \mathcal{N}_f$. *Then*

$$(\mathcal{N}_f^{(i)}h) \circ f^{(i)}(X) = \prod_{\omega \in W_f^i} h(X[+]\omega).$$

*(iv) If* $h \in R$, $h \equiv 1 \bmod \wp'^i$ $(i \geq 1)$, *then* $\mathcal{N}h \equiv 1 \bmod \wp'^{i+1}$.

PROOF: Clearly (1) characterizes the power series $\mathcal{N}h$ uniquely, and from this $\mathcal{N}(h_1h_2) = \mathcal{N}h_1 \cdot \mathcal{N}h_2$, if we only show how to construct $\mathcal{N}h$. Let $g_0(X)$ be the right hand side in (1). Then $g_0(X[+]\omega) = g_0(X)$ $(\omega \in W_f^1)$ so $g_0(X) - g_0(0) = g_1(X) \cdot f(X)$, by Weierstrass preparation theorem. Now $g_1(X)$ is $W_f^1$-translation-invariant, so similarly $g_1(X) - g_1(0) = g_2(X) \cdot f(X)$, etc. Hence

$$g_0(X) = g_0(0) + g_1(0)f(X) + g_2(0)f(X)^2 + \cdots$$

and this infinite series converges in the topology of $R$. $\mathcal{N}h = g_0(0) + g_1(0)X + g_2(0)X^2 + \cdots$ therefore satisfies (1). To prove (i) note that

$$\mathcal{N}h(X^q) \equiv \mathcal{N}h \circ f \equiv h(X)^q \equiv h^{\varphi}(X^q) \bmod \wp'.$$

Point (ii) is obvious, applying $\varphi$ to (1). The case $i = 1$ of (iii) is (1) and to prove the general case we proceed by induction, assuming it for $i - 1$. Then

$$\begin{aligned}
\mathcal{N}_f^{(i)} h \circ f^{(i)}(X) &= \mathcal{N}_{\varphi(f)}^{(i-1)}(\mathcal{N}_f h) \circ \varphi(f)^{(i-1)}(f(X)) \\
&= \prod_{\beta \in W^{i-1}_{\varphi(f)}} \mathcal{N}_f h(f(X)[+]_{\varphi(f)}\beta) \qquad \text{(induction)} \\
&= \prod_{\alpha \in W^i_f \ mod\ W^1_f} \mathcal{N}_f h(f(X[+]_f \alpha)) \qquad (f : W^i_f \to W^{i-1}_{\varphi(f)}) \\
&= \prod_{\alpha \in W^i_f \ mod\ W^1_f} \prod_{\gamma \in W^1_f} h(X[+]_f \alpha [+]_f \gamma) \qquad \text{(by (1))} \\
&= \prod_{\omega \in W^i_f} h(X[+]_f \omega) \qquad \text{(grouping together).}
\end{aligned}$$

Finally let $\wp_n$ be the valuation ideal in $k^n_\xi$ (so $\wp_0 = \wp'$). We prove (iv) by induction. What we need from the $i - 1$ step is that if $h \equiv 1 \, mod \, \wp'^i$, then $\mathcal{N} h \equiv 1 \, mod \, \wp'^i$. For $i = 1$ this is easy to check directly. Now let $i \geq 1$ and consider the congruence

$$\mathcal{N} h(X^q) \equiv \mathcal{N} h \circ f \equiv h(X)^q \equiv 1 \, mod \, \wp'^i \wp_1$$

which holds because $h \equiv \mathcal{N} h \equiv 1 \, mod \, \wp'^i$. Since $\mathcal{N} h \in \mathcal{O}'[[X]]$, actually $\mathcal{N} h \equiv 1 \, mod \, \wp'^{i+1}$.

Call $\mathcal{N}$ (Coleman's) *norm operator*. Notice that if $h \in X^i R^x$ $(i \geq 0)$ then $\mathcal{N} h \in X^i R^x$ as well, so $\mathcal{N}$ extends to $\mathcal{O}'((X))^x$ ($\mathcal{O}'((X))$ is the ring of Laurent series over $\mathcal{O}'$) and the same remark holds there too. We shall be applying $\mathcal{N}$ primarily to power series in $\mathcal{O}'((X))^x$. If $a$ is in $\mathcal{O}'$, $\mathcal{N} a = a^q$.

**2.2 Theorem.** *Let $\beta = (\beta_n)$ be a norm-coherent sequence in $(k^n_\xi)$ (i.e. $\beta_n \in (k^n_\xi)^x$ for $n \geq 0$, and $N_{m,n}(\beta_m) = \beta_n$ where $N_{m,n}$ is the norm from $k^m_\xi$ to $k^n_\xi$ for $m \geq n$). Let $\nu(\beta)$ be the normalized valuation of $\beta$ (i.e. $\beta_n \mathcal{O}_n = \wp_n^{-\nu(\beta)}$ for any $n \geq 0$, where $\mathcal{O}_n$ and $\wp_n$ are the valuation ring and ideal in $k^n_\xi$). Fix $f \in \mathcal{F}_\xi$, and $\omega_i \in W^i_{\varphi^{-i}(f)} \backslash W^{i-1}_{\varphi^{-i}(f)} = \tilde{W}^i_{\varphi^{-i}(f)}$ such that $(\varphi^{-i} f)(\omega_i) = \omega_{i-1}$ $(1 \leq i < \infty)$. Then there exists a unique $g_\beta \in X^{\nu(\beta)} \cdot \mathcal{O}'[[X]]^x$ such that $(\varphi^{-i} g_\beta)(\omega_i) = \beta_i$ for all $i \geq 1$.*

By abuse of language we shall call $(\omega_i)_{i \geq 0}$ a *generator of the Tate module* of $F_f$ (but notice that $\omega_i$ is a division point on $F_{\varphi^{-i}(f)}$ and not on $F_f$). We shall call

$g_\beta(X)$ *Coleman's power series* of $\beta$. It converts $f$-compatible division points into norm-compatible elements. A stronger result holds "at finite levels $n$" but we shall not need it, and its proof is slightly more complicated (see [Col1] for the "absolute" case $k' = k$).

PROOF: Fix $m \geq 1$. Since $\omega_m$ is a prime of $k_\xi^m$, which is a totally ramified extension of $k'$, there exists $h \in X^{\nu(\beta)}R^x$ such that $h(\omega_m) = \beta_m$. If $1 \leq n \leq m$,

$$(2) \qquad \begin{aligned} (\mathcal{N}_{\varphi^{-m}f}^{(m-n)}h)(\omega_n) &= \mathcal{N}_{\varphi^{-m}f}^{(m-n)}h \circ (\varphi^{-m}f)^{(m-n)}(\omega_m) \\ &= \prod_{\alpha \in W_{\varphi^{-m}f}^{m-n}} h(\omega_m[+]\alpha) \\ &= N_{m,n}(h(\omega_m)) = \beta_n. \end{aligned}$$

On the other hand,

$$(3) \qquad \frac{\mathcal{N}_{\varphi^{-m}f}^{(m)}h}{\varphi^n\left(\mathcal{N}_{\varphi^{-m}f}^{(m-n)}h\right)} = \mathcal{N}_{\varphi^{n-m}f}^{(m-n)}\left(\frac{\mathcal{N}_{\varphi^{-m}f}^{(n)}h}{\varphi^n h}\right) \equiv 1 \ mod \ \wp'^{m-n+1}.$$

The first equality follows from multiplicativity and prop. 2.1(ii). Then from the fact that $h \in X^{\nu(\beta)}R^x$ and from 2.1(i) the quantity in parenthesis is in $1 + \wp' R$. The congruence now follows by successive application of 2.1(iv). Let $g_m = \mathcal{N}_{\varphi^{-m}f}^{(m)}h$. Together (2) and (3) imply for all $1 \leq n \leq m$ that $(\varphi^{-n}g_m)(\omega_n)/\beta_n \equiv 1 \ mod \ \wp'^{m-n+1}$. Since $X^{\nu(\beta)}R^x$ is compact, we may choose in it a limit point $g$ of $\{g_m\}$. Then by continuity $(\varphi^{-n}g)(\omega_n) = \beta_n$ for all $1 \leq n$. The Weierstrass preparation theorem shows that $g = g_\beta$ is unique, so actually $g = \lim_{m\to\infty} g_m$. This concludes the proof of the theorem.

**2.3 Corollary.**

*(i)* $g_{\beta\beta'} = g_\beta \cdot g_{\beta'}$,

*(ii)* $\mathcal{N}_f g_\beta = g_\beta^\varphi$,

*(iii)* $g_\beta(0)^{1-\varphi^{-1}} = \beta_0 \qquad (\nu(\beta) = 0)$.

*(iv) If* $\sigma \in Gal(k_\xi/k')$ *and* $\kappa(\sigma) \in \mathcal{O}^x$ *is the unique unit for which* $\sigma(\omega) = [\kappa(\sigma)]_f(\omega)$ *for any* $\omega \in W_f$ *(and any* $f \in \mathcal{F}_\xi$*—see 1.8;* $k_\xi = \cup k_\xi^n$, $W_f = \cup W_f^n$*), then* $g_{\sigma(\beta)} = g_\beta \circ [\kappa(\sigma)]_f$.

PROOF: Point (i) is clear from the defining property of $g_\beta$. For (ii) notice that

$$\begin{aligned}\beta_{i-1} &= N_{i,i-1}\beta_i = \prod_{\alpha\in W^1_{\varphi^{-i}f}} (\varphi^{-i}g_\beta)(\omega_i[+]_{\varphi^{-i}f}\alpha) \\ &= \mathcal{N}_{\varphi^{-i}f}(\varphi^{-i}g_\beta) \circ (\varphi^{-i}f)(\omega_i) \\ &= \varphi^{-i}(\mathcal{N}_f g_\beta)(\omega_{i-1}),\end{aligned}$$

while we also have $\beta_{i-1} = (\varphi^{1-i}g_\beta)(\omega_{i-1})$. Hence $g_\beta^\varphi - \mathcal{N}_f g_\beta$ has infinitely many zeroes, so must be identically 0. Point (iii) follows from (ii) because when $\nu(\beta) = 0$, $g_\beta(0) = \varphi^{-1}(\mathcal{N}_f g_\beta)(0) = \mathcal{N}_{\varphi^{-1}f}(\varphi^{-1}g_\beta)(\varphi^{-1}f(0)) = \prod_{\alpha\in W^1_{\varphi^{-1}f}}(\varphi^{-1}g_\beta)(\alpha) = \varphi^{-1}g_\beta(0) \cdot N_{1,0}\beta_1 = \varphi^{-1}g_\beta(0) \cdot \beta_0$. Note that (iii) is in accordance with the known fact that $N_{k'/k}(\beta_0) \in <\xi>$, and if $\beta$ is a unit, $N_{k'/k}(\beta_0) = 1$. Finally (iv) follows from $\varphi^{-i}(g_\beta \circ [\kappa(\sigma)]_f)(\omega_i) = (\varphi^{-i}g_\beta) \circ [\kappa(\sigma)]_{\varphi^{-i}f}(\omega_i) = (\varphi^{-i}g_\beta)(\sigma\omega_i) = \sigma((\varphi^{-i}g_\beta)(\omega_i)) = \sigma\beta_i$.

For a generalization of 2.3(iv) to $\sigma \in Gal(k_\xi/k)$ see 3.7(15).

## 3. MEASURES FROM UNITS

It is instructive to think of (classical, abelian) $L$-functions as functions of characters of the Galois group, or quasi-characters of the idele group. With this in mind, $\mathrm{N}\mathfrak{p}^{-s}$ and $\chi(\mathfrak{p})$ appear on equal footing, because both arise from characters on the idele group.

$p$-adic (abelian) $L$-functions, like those of Kubota-Leopoldt, Deligne-Ribet, or Katz, should be conceived in the same light. This time the situation is even better, because grossencharacters of type $A_0$ (and in particular $\mathbf{N}$) which could not be interpreted as $\mathbf{C}$-valued characters of any Galois group, become just so when considered $p$-adically, i.e. as $\mathbf{C}_p$-valued ($\mathbf{C}_p$ is the completion of an algebraic closure of $\mathbf{Q}_p$). Needless to say, the Galois group is now *profinite* and not necessarily finite. This point of view is due to Weil [We]. As functions of $p$-adic characters on a group $G$, the $p$-adic $L$ functions are more than just locally analytic. They belong to the *Iwasawa algebra*. This means that their value at a character $\chi$ is obtained by integrating $\chi$ against a *p-adic measure* on G. For all practical purposes, the $p$-adic $L$ function may be identified with the measure. See [Maz1] and [Se2]. The translation into the language of power series, or functions of "$s$", is routine, as will be explained, and not always beneficial.

In this section we first review some general definitions. We then go back to the situation of §2, but assume that the formal group is of height 1 ($k = \mathbf{Q}_p$). The method of Coleman's power series is used to turn a coherent sequence of units into a measure on the local Galois group. This important procedure underlines the construction of chapter II.

**3.1** Let $M$ be an abelian group and $G$ a profinite group (usually a Galois group). An $M$-valued *distribution* on $G$ is a finitely additive function from the Boolean algebra of compact-open subsets of $G$ to $M$. We denote the abelian group of $M$-valued distributions on $G$ by $\Lambda(G, M)$. If $M$ is an $A$-algebra for a commutative ring $A$, so is $\Lambda(G, M)$, and $\Lambda(G, A)$ is even a *ring* with *convolution* as product. The convolution of $\lambda$ and $\mu$ is defined as follows. If $U$ is open and compact, so is the subgroup $H = \{\gamma \in G \mid U\gamma = U\}$, so $G = \cup_{i=1}^{n} \sigma_i H$, a disjoint union. Set $(\lambda \cdot \mu)(U) = \sum_{i=1}^{n} \lambda(U\sigma_i^{-1}) \cdot \mu(\sigma_i H)$ $(= \int_G \lambda(U\sigma^{-1})d\mu(\sigma))$.

Now suppose $M \subset \mathbf{C}_p$. If $M$ is bounded ($|x| \leq R < \infty$ for all $x \in M$ and some $R$) we call an $M$-valued distribution a *p-adic measure.* If $M \subset \mathbf{D}_p = \{x \in \mathbf{C}_p \mid |x| \leq 1\}$ we talk about *integral measures.*

If $G$ is finite $\Lambda(G, M) \simeq M[G]$ under $\lambda \mapsto \sum_{\sigma \in G} \lambda(\{\sigma\})\sigma$. If $M = A$ is a commutative ring, convolution corresponds to the usual product. In general, $\Lambda(G, M) = \varprojlim \Lambda(G/H, M) \simeq \varprojlim M[G/H] = M[[G]]$ (notation), where the inverse limit is over the family of normal subgroups $H$ of finite index in $G$. $M[G]$ is dense in $M[[G]]$.

The *(standard) Iwasawa algebra* is obtained when $M = \mathbf{Z}_p$ and $G = \Gamma$, a group isomorphic to $\mathbf{Z}_p$ (usually $1 + p\mathbf{Z}_p$ for odd $p$, $1 + 4\mathbf{Z}_2$ when $p = 2$). It is well known that in this case $\Lambda \simeq \mathbf{Z}_p[[S]]$ non-canonically. The isomorphism depends on a choice of a topological generator $u$ of $\Gamma$, and maps $u^\alpha$ to $(1 + S)^\alpha$.

When $\Gamma = \mathbf{Z}_p$ however (written additively), we naturally take $u = 1$. The power series $P_\mu(S)$ corresponding to $\mu$ is then

$$P_\mu(S) = \int_{\mathbf{Z}_p} (1 + S)^\alpha \, d\mu(\alpha). \tag{1}$$

If $\chi : G \to \mathbf{C}_p$ is any continuous function, and $\lambda$ is a $p$-adic measure, then the *Riemann integral* $\int_G \chi(\sigma) d\lambda(\sigma)$ exists. Simply approximate $\chi$ uniformly by locally constant functions. In particular, if $\chi \in Hom(G, \mathbf{C}_p^\times)$, then

$$\int_G \chi \, d(\lambda\mu) = \int_G \chi \, d\lambda \cdot \int_G \chi \, d\mu. \tag{2}$$

The *augmentation ideal* in $\Lambda(G, A)$ is the kernel of $\mu \mapsto \mu(G)$.

Finally, assume that $G$ is commutative, and let $S \subset \Lambda(G, A)$ be the multiplicative set of non-zero-divisors. A *pseudo-measure* is an element $\lambda/s$ of $S^{-1}\Lambda(G, A)$. If $\chi \in Hom(G, \mathbf{C}_p^\times)$ and $\int \chi \, ds \neq 0$, set $\int \chi \, d(\lambda/s) = \int \chi \, d\lambda / \int \chi \, ds$. In view of (2), this is well defined.

REMARK: If $G$ admits subgroups whose indices are divisible by arbitrarily high powers of $p$, then there is no non-trivial $p$-adic measure on $G$ which is translation invariant. In our framework, therefore, there are no Haar measures.

**3.2** A FORMAL CONSTRUCTION OF MEASURES: We return to the situation of §§1,2 and assume from now on that $F_f$ is a relative Lubin-Tate group *of height one.* Thus $k = \mathbf{Q}_p$, $\mathcal{O} = \mathbf{Z}_p$, and $k'$ is an unramified extension of $\mathbf{Q}_p$. By proposition 1.6 there exists an isomorphism

$$\theta : \hat{\mathbf{G}}_m \simeq F_f, \quad T = \theta(S) = \Omega \cdot S + \cdots \in \mathcal{O}_K[[S]], \tag{3}$$

over the ring of integers of the completion of $k^{ur}$. Let $f(T) = \pi' T + \cdots$ be the special endomorphism of $F_f$. The special endomorphism of $\hat{\mathbf{G}}_m$ is of course $[p](S) = pS + \cdots$. Proposition 1.6 implies then

$$\Omega^{\varphi - 1} = \pi'/p, \quad f \circ \theta = \theta^\varphi \circ [p], \tag{4}$$

where $\varphi$ is, as usual, the Frobenius automorphism.

Fix once and for all primitive $p^n$ roots of unity $\varsigma_n$ $(n \geq 0)$ such that $\varsigma_n^p = \varsigma_{n-1}$. Letting

$$(5) \qquad \omega_n = \theta^{\varphi^{-n}} (\varsigma_n - 1)$$

we see that $\omega_n \in \tilde{W}^n_{\varphi^{-n}f}$ and $(\varphi^{-n}f)(\omega_n) = \omega_{n-1}$. Thus $(\omega_n)$ is a generator of the Tate module of $F_f$ in the sense of 2.2.

Let $\lambda = \lambda_f$ be the logarithm of $F_f$, normalized so that $\lambda'(0) = 1$. Then $\lambda'(T) \in \mathcal{O}'[[T]]^x$ (cf. 1.1).

**3.3** For any local field $k$ let $U(k)$ be the subgroup of *principal units* (units congruent to 1 *modulo* $\wp$). Let

$$(6) \qquad \beta = (\beta_n) \in \mathcal{U} = \lim_{\leftarrow} U(k^n_\xi)$$

be a norm coherent sequence of principal units in the tower $k^n_\xi = k'(W^n_f)$. In theorem 2.2 we attached to $\beta$ a power series $g_\beta(T) \in \mathcal{O}'[[T]]^x$ such that $(\varphi^{-n}g_\beta)(\omega_n) = \beta_n$ $(n \geq 1)$. Since $\beta_n$ are principal units $g_\beta \equiv 1$ *mod* $(\wp', T)$ and we can take its logarithm using the power series expansion of log $(1 + h)$. Thus $\log\, g_\beta \in k'[[T]]$.

**Lemma**. *The power series*

$$(7) \qquad \widetilde{\log}\, g_\beta(T) = \log\, g_\beta(T) - \frac{1}{p} \sum_{\omega \in W^1_f} \log\, g_\beta(T[+]\omega)$$

*has integral coefficients.*

PROOF: Let $g = g_\beta$. Since $g^p \equiv g^\varphi \circ f$ *mod* $\wp'$ we get from corollary 2.3(ii) that $g^p \equiv \prod g(T[+]\omega)$ *mod* $\wp'$. Taking logarithms, and noting that $np \mid p^n$ for $n \geq 1$, we see that $p$ times the right hand side of (7) is congruent to 0 *mod* $\wp'$, which proves the lemma.

We remark that (7) is also equal to $\log\, g_\beta - \dfrac{1}{p} \log(g^\varphi_\beta \circ f)$.

Let $\tilde{a}_\beta(S) = \widetilde{\log g_\beta} \circ \theta(S) \in \mathcal{O}_K[[S]]$ and let $\mu_\beta$ be the $\mathcal{O}_K$-valued measure on $\mathbf{Z}_p$ for which $P_\mu = \tilde{a}_\beta$. In other words,

$$(8) \qquad \tilde{a}_\beta(S) = \int_{\mathbf{Z}_p} (1 + S)^\alpha \, d\mu_\beta(\alpha).$$

The measure $\mu_\beta$ is actually supported on $\mathbf{Z}_p^x$. This is a consequence of (7)—removing the Euler factor at $p$—and without it we would get a distribution, but not a measure. Indeed, if $\tilde{\mu} = \mu|\mathbf{Z}_p^x$, extended by 0 to $p\mathbf{Z}_p$, then $P_{\tilde{\mu}} = \tilde{P}_\mu$, where

$$(7') \qquad \tilde{P}_\mu(S) = P_\mu(S) - \frac{1}{p} \sum_{\varsigma^p = 1} P_\mu(\varsigma(1 + S) - 1).$$

But with $P_\mu = \tilde{a}_\beta$, (7) implies $\tilde{P}_\mu = P_\mu$, hence $\tilde{\mu} = \mu$.

We may now use the isomorphism

$$(9) \qquad \begin{cases} \kappa : G \simeq \mathbf{Z}_p^x & G = Gal(k_\xi/k') \\ \sigma(\omega) = [\kappa(\sigma)]_f(\omega) & \omega \in W_f \end{cases}$$

to pull back $\mu_\beta$ to $G$ (cf. 2.3(iv). Note that when we follow $\kappa$ by the local Artin symbol—see 1.8—$\sigma$ goes to $\sigma^{-1}$). We still denote our measure by $\mu_\beta$.

**3.4** DEFINITION. *For any $\beta \in \mathcal{U}$ let $\mu_\beta$ be the $\mathcal{O}_K$ – valued measure on $G = Gal(k_\xi/k')$ satisfying*

$$(10) \qquad \widetilde{\log g_\beta} \circ \theta(S) = \int_G (1 + S)^{\kappa(\sigma)} \, d\mu_\beta(\sigma).$$

Some elementary properties of the map $\beta \mapsto \mu_\beta$ are summarized below.

**Lemma.**

*(i)* $\mu_{\beta\beta'} = \mu_\beta + \mu_{\beta}$.

*(ii) Let $\gamma \in G$. Then $\mu_{\gamma(\beta)}(\gamma U) = \mu_\beta(U)$.*

*(iii) $\mu_\beta$ depends only on the choice of $(\varsigma_n)$. If $\varsigma_n' = \varsigma_n^{\kappa(\gamma)}$ $(\gamma \in G)$, then the resulting measure is given by $\mu_\beta'(U) = \mu_\beta(\gamma U)$.*

PROOF: (i) and (ii) are consequences of 2.3(i) and (iv) respectively, and (10). Part (iii) is proved similarly, and is left to the reader.

Let $\mathcal{G} = Gal(k_\xi/k)$, so that $\mathcal{G}/G = Gal(k'/k)$ is cyclic of order $d$. Suppose $U$ is an open subset of $\mathcal{G}$ contained in a coset of $G$. If $\gamma U \subset G$ define $\mu_\beta(U) = \mu_{\gamma(\beta)}(\gamma U)$ (now $\gamma \in \mathcal{G}$). Part (ii) of the lemma shows that this is independent of $\gamma$. We can now *extend* $\mu_\beta$ *to a measure on* $\mathcal{G}$, so we get a map $i: \mathcal{U} \to \Lambda(\mathcal{G}, \mathcal{O}_K)$, $i(\beta) = \mu_\beta$. Since $g_\beta$ is non-constant for $\beta \neq 1$, (10) shows that $i$ is injective. The lemma implies the following.

**Corollary.** *The map $i$ is an injective homomorphism of $\mathbf{Z}_p[[\mathcal{G}]]$ – modules.*

**3.5** THE COATES-WILES HOMOMORPHISMS: These variants of $\mu_\beta$ were first considered by Kummer ([Kum] p. 493) and are also known as *Kummer's logarithmic derivatives.* Let $D = \dfrac{\Omega}{\lambda'(T)} \dfrac{d}{dT}$ be the translation invariant derivation of $F_f$ over $\mathcal{O}_K$ ($\Omega$ as in (3)). Letting $T = \theta(S)$ and using $\lambda \circ \theta(S) = \Omega \cdot \log(1 + S)$ we see that in terms of $S$, $D = (1 + S)\dfrac{d}{dS}$, the standard translation invariant derivation of $\hat{\mathbf{G}}_m$. The *moments* of $\mu_\beta$ are given by the formula

$$\int_G \kappa(\sigma)^k d\mu_\beta(\sigma) = D^k \, \widetilde{\log g_\beta}(0) \quad (k \geq 0). \tag{11}$$

When $F_f$ is an (absolute) Lubin Tate group, so $\varphi$ acts trivially on $g_\beta(T)$, this can also be written $\left(1 - \dfrac{\pi^k}{p}\right) D^k \log g_\beta(0)$, $\pi = f'(0)$. The map $\varphi_k(\beta) = \int_G \kappa(\sigma)^k d\mu_\beta(\sigma)$ is called the $k^{th}$ Coates-Wiles homomorphism. It satisfies

(i) $\varphi_k(\beta\beta') = \varphi_k(\beta) + \varphi_k(\beta')$

(ii) $\varphi_k(\gamma(\beta)) = \kappa(\gamma)^k \cdot \varphi_k(\beta) \quad (\gamma \in G)$.

Together $\varphi_k$ for $k \geq 0$ determine $\mu_\beta$ uniquely on $G$ (but not on $\mathcal{G}$).

**3.6** Another useful formula is for $\mu_\beta(G_n)$, where $G_n = Gal(k_\xi/k_\xi^n) = \kappa^{-1}(1 + p^n\mathbf{Z}_p)$, $n \geq 1$. From (8) we deduce at once

$$\mu_\beta(G_n) = \frac{1}{p^n} \sum_{j=0}^{p^n-1} \tilde{a}_\beta(\zeta_n^j - 1) \cdot \zeta_n^{-j}. \tag{12}$$

Here $\zeta_n$ can be *any* primitive $p^n$ root of 1. The formula remains valid with $a_\beta$ instead of $\tilde{a}_\beta$.

**3.7** Let $N \geq 0$ be the largest integer such that $\zeta_N \in k_\xi$. If $[k' : \mathbf{Q}_p] = d$, local class field theory shows that $N$ is the largest integer for which $p^d \xi^{-1} \equiv 1 \; mod \; p^N$ (cf. the discussion at the end of 1.8). $N = \infty$ if and only if $F_f \cong \hat{\mathbf{G}}_m$ over $\mathcal{O}'$. We call $N$ the *anomality index* of $F_f$ (over $k'$), and call $F_f$ itself anomalous if $N > 0$. Since $\mathcal{G}$ acts on $\mu_{p^N}$, $\mathcal{O}_K \otimes \mu_{p^N} = \mathcal{O}_K/p^N(1)$ (Tate twist) is a $\Lambda(\mathcal{G}, \mathcal{O}_K)$-module.

Let $\mathbf{N} : \mathcal{G} \to (\mathbf{Z}/p^N\mathbf{Z})^x$ be the character giving the action on $\mu_{p^N}$. Observe that for $\gamma \in G$, $\mathbf{N}(\gamma) \equiv \kappa(\gamma) \; mod \; p^N$. Define a map $j : \Lambda(\mathcal{G}, \mathcal{O}_K) \to \mathcal{O}_K/p^N(1)$ by

$$j(\mu) = \int_{\mathcal{G}} \mathbf{N}(\sigma) d\mu(\sigma).$$

Then $j$ is a surjective homomorphism of $\mathcal{G}$ modules. In the following key result $i$ was extended by linearity to the completed tensor product $\mathcal{U} \hat{\otimes}_{\mathbf{Z}_p} \mathcal{O}_K$. Its proof will take up the rest of this section.

**Theorem.** *The sequence*

$$0 \to \mathcal{U} \hat{\otimes}_{\mathbf{Z}_p} \mathcal{O}_K \xrightarrow{i} \Lambda(\mathcal{G}, \mathcal{O}_K) \xrightarrow{j} (\mathcal{O}_K/p^N)(1) \to 0 \tag{13}$$

*is exact. (If $N = \infty$ it ends with $\mathcal{O}_K(1)$).*

PROOF: We have already mentioned that $j$ is surjective. We shall show now that $j \circ i = 0$. Let $\beta \in \mathcal{U}$. A calculation based on (11) gives

$$\int_G \mathbf{N}(\sigma) d\mu_\beta(\sigma) \equiv \int_G \kappa(\sigma) d\mu_\beta(\sigma) \equiv \Omega \frac{g'_\beta(0)}{g_\beta(0)} - \left(\Omega \frac{g'_\beta(0)}{g_\beta(0)}\right)^\varphi \; mod \; p^N.$$

Now $\kappa$ can be extended to $\mathcal{G}$ in the following way. If $\sigma \in \mathcal{G}$, there is a unique isomorphism $h : F_f \simeq F_{\sigma(f)}$ such that $h(\omega) = \sigma(\omega)$ for all $\omega \in W_f$. This $h$ is of the form $[\kappa(\sigma)]_{f,\sigma(f)}$ for a unique $\kappa(\sigma) \in \mathcal{O}'^x$, and $\kappa(\sigma)^{\varphi - 1} = f'(0)^{\sigma - 1}$. Then $\kappa : \mathcal{G} \to (\mathcal{O}')^x$ is a 1-cocycle

$$\kappa(\tau\sigma) = \kappa(\sigma)^\tau \cdot \kappa(\tau), \tag{14}$$

and the generalization of corollary 2.3(iv) is

$$g_{\sigma(\beta)} = g_\beta^\sigma \circ [\kappa(\sigma)]_{f,\sigma(f)}. \tag{15}$$

From the definition of $\mu_\beta$ one gets now

$$\int_{\mathcal{G}} \mathrm{N}(\sigma) d\mu_\beta(\sigma) \equiv \sum_{i=0}^{d-1} \varphi^i(1 - \varphi)\left(\Omega \frac{g'_\beta(0)}{g_\beta(0)}\right)$$
$$\equiv \Omega \frac{g'_\beta(0)}{g_\beta(0)} \cdot \left(1 - \frac{\xi}{p^d}\right) \equiv 0 \; mod \; p^N.$$

This proves $j \circ i = 0$. The rest is more difficult. Although not absolutely necessary, we shall first reduce the proof to the case of $\hat{\mathbf{G}}_m$ (sections 3.8-3.9).

**3.8** We want to change perspective somewhat. The local Artin map identifies $\mathcal{U} = \varprojlim U(k_\xi^n)$ with $Gal(k^{ur}M(k_\xi)/k^{ab})$ where $M(L)$, for any $L$, is the *maximal abelian p − extension* of $L$. Viewing $F_f$ over an unramified extension $k''$ of $k'$, and letting $\mathcal{G}' = Gal(k''k_\xi/k)$, $\mathcal{U}' = \varprojlim U(k''k_\xi^n)$ (note that $k''k_\xi^n = k_{\xi'}^n$, with $\xi' = \xi^{[k'':k']}$), we obtain an exact sequence like (13), but over $k''$. The two exact sequences are related through a commutative diagram.

(16)
$$\begin{array}{ccccccccc} 0 \to & \mathcal{U}'\hat{\otimes}\mathcal{O}_K & \xrightarrow{i'} & \Lambda(\mathcal{G}',\mathcal{O}_K) & \xrightarrow{j'} & (\mathcal{O}_K/p^{N'})(1) & \to 0 \\ & \downarrow & & \downarrow & & \downarrow & \\ 0 \to & \mathcal{U}\hat{\otimes}\mathcal{O}_K & \xrightarrow{i} & \Lambda(\mathcal{G},\mathcal{O}_K) & \xrightarrow{j} & (\mathcal{O}_K/p^{N})(1) & \to 0. \end{array}$$

The first vertical arrow is $N_{k''/k'} \otimes 1$, the middle one is the map induced from the projection $\mathcal{G}' \to \mathcal{G}$ and the third is the canonical map. All three are surjective. Taking the projective limit over $k' \subset k'' \subset k^{ur}$ we arive at

(17)
$$0 \to Gal(M(k^{ab})/k^{ab}) \hat{\otimes} \mathcal{O}_K \xrightarrow{i} \Lambda(\mathcal{G}_a, \mathcal{O}_K) \xrightarrow{j} \mathcal{O}_K(1) \to 0$$

where $\mathcal{G}_a = Gal(k^{ab}/k)$, and where we used the Artin map to identity $\varprojlim \mathcal{U}'$ with the Galois group.

**Lemma**. *(17) is exact if and only if (13) is (for every $k'$).*

PROOF: If (13) is exact for every $k'$, then so is (17) because the vertical arrows in (16) are surjective. Suppose (17) is exact. Let $H = Gal(k^{ab}/k_\xi) \cong Gal(k^{ur}/k')$ and let $\alpha$ be the generator of $H$ restricting to $\varphi^d$ on $k^{ur}$. Then (13) is the sequence of $\alpha$-coinvariants of (17) (the $\alpha$-coinvariants of a module $E$ are $E/(\alpha - 1)E$). If

$(\alpha - 1) \mid \mathcal{O}_K(1)$ is injective, the snake lemma concludes the proof. This is always the case, unless $F_f \simeq \hat{\mathbf{G}}_m$ over $\mathcal{O}'$. Since in the case of $\hat{\mathbf{G}}_m$ we shall prove the theorem directly, we assume that $F_f \not\simeq \hat{\mathbf{G}}_m$ over $\mathcal{O}'$.

The advantage of (17) is that the particular formal group disappears from the scene. The following proposition is interesting in its own right.

**3.9 Proposition.** *The homomorphisms $i$ and $j$ in (17) are independent of $k'$, $\xi$ or $f \in \mathcal{F}_\xi$. In other words, they are canonically associated to $k = \mathbf{Q}_p$.*

PROOF: Let $F_1$ and $F_2$ be two relative Lubin Tate groups, both defined over $\mathcal{O}'$. Let $\tau \in Gal(M(k^{ab})/k^{ab})$ and let $i_1(\tau) = \mu_1$ and $i_2(\tau) = \mu_2$ be the measures associated to $\tau$ on $\mathcal{G}_a$ (relative to $F_1$ and $F_2$). Let $\theta_1 : \hat{\mathbf{G}}_m \simeq F_1$ and $\eta : F_1 \simeq F_2$ be isomorphisms over $\mathcal{O}_K$, $\theta_2 = \eta \circ \theta_1$, and set $\omega_{1,n} = (\varphi^{-n}\theta_1)(\varsigma_n - 1)$ and similarly $\omega_{2,n}$, as in (5) above. Write $k_1^n = k'(\omega_{1,n})$ and similarly $k_2^n$ for the fields denoted $k_\xi^n$ before. For each $n$ there exists an unramified extension $k''$ of $k'$ for which $k''k_1^n = k''k_2^n$ (see 1.8) and we consider the collection $\mathcal{S}$ of all subgroups of $\mathcal{G}_a$ of the form $H = Gal(k^{ab}/k''k_1^n)$ for such pairs $(n, k'')$. $\mathcal{S}$ is a basis of neighborhoods at zero, and we shall prove $\mu_1(H) = \mu_2(H)$ for $H \in \mathcal{S}$. This, for all $\tau$, will prove our claim, because

$$(18) \qquad i(\tau)(\sigma H) = i(\tilde{\sigma}^{-1}\tau\tilde{\sigma})(H)$$

where $\sigma \in \mathcal{G}_a$ and $\tilde{\sigma}$ is any extension to $M(k^{ab})$, and where $i$ is $i_1$ or $i_2$ (see lemma 3.4(ii)).

So fix $n$ and $k''$ as above and consider $F_1$ and $F_2$ over $k''$. Let $\beta_1 = (\beta_{1,m}) \in \mathcal{U}_1 = \varprojlim U(k''k_1^m)$ and (similarly) $\beta_2 \in \mathcal{U}_2$ correspond to $\tau$. Then $\beta_{1,m} = \beta_{2,m}$ for $0 \leq m \leq n$. The field diagram is as below.

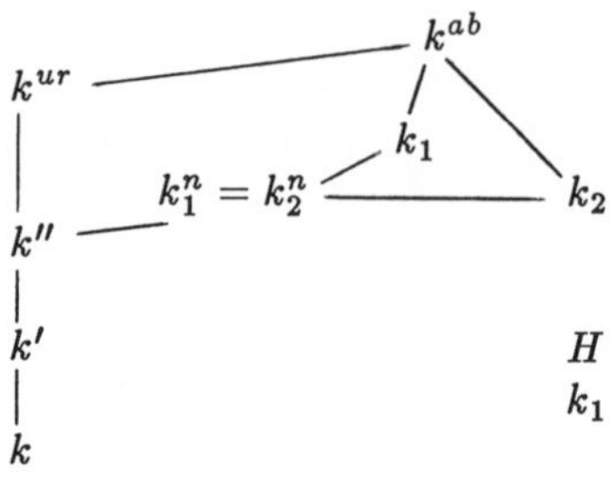


$$H = Gal(k^{ab}/k_1^n)$$
$$k_1 = \cup k_1^m, \quad k_2 = \cup k_2^m.$$

Let $g_1$ and $g_2$ be the Coleman power series of $\beta_1$ and $\beta_2$ (relative to $F_1$ and $F_2$). From 3.6 (12) we get

$$\mu_1(H) = \frac{1}{p^n} \sum_{j=0}^{p^n-1} \log g_1 \circ \theta_1(\varsigma_n^j - 1) \cdot \varsigma_n^{-j} \tag{19}$$

and similarly for $\mu_2(H)$. Now for $0 < m \leq n$

$$\varphi^{-m}(g_1 \circ \theta_1)(\varsigma_m - 1) = \beta_{1,m} = \beta_{2,m} = \varphi^{-m}(g_2 \circ \theta_2)(\varsigma_m - 1),$$

so extending $\varphi$ to $k^{ab}$ so as to fix all $\varsigma_m$, and applying $\varphi^m$ to the last equality, we see that $g_1 \circ \theta_1(\varsigma_m - 1) = g_2 \circ \theta_2(\varsigma_m - 1)$ for all $0 < m \leq n$. Conjugating by $Gal(K(\varsigma_m)/K)$ it follows that $g_1 \circ \theta_1(\varsigma_n^j - 1) = g_2 \circ \theta_2(\varsigma_n^j - 1)$ for all $0 < j \leq p^n - 1$. It remains to show $g_1(0) = g_2(0)$. Note first that $X^{-1}\{(1 + X)^{p^n} - 1\}$ divides $g_1 \circ \theta_1 - g_2 \circ \theta_2$, so $g_1(0) \equiv g_2(0)$ *mod* $p^n$. Consider now $F_1$ and $F_2$ *over* $k'$, let $b_1 = N_{k''/k'}\beta_1$, $b_2 = N_{k''/k'}\beta_2$ be the inverse systems of local units corresponding to $\tau$ over $k'$, and $h_1 = N_{k''/k'}g_1$, $h_2 = N_{k''/k'}g_2$ their Coleman power series. It follows that $h_1(0) \equiv h_2(0)$ *mod* $p^n$. With $k'$ held fixed, $b_i$ and $h_i$ are fixed, and $n$ can be chosen arbitrarily large. Hence $h_1(0) = h_2(0)$. Repeating this argument with $k''$ instead of $k'$, we deduce $g_1(0) = g_2(0)$, and the proof of the proposition is concluded.

**3.10** Proposition 3.9 shows that it is enough to prove our theorem in the case $F_f = \hat{\mathbf{G}}_m$ (over arbitrary $k'$). In this case we shall actually show that

$$(20) \qquad 0 \to \mathcal{U} \otimes_{\mathbf{Z}_p} \mathcal{O}' \xrightarrow{i} \mathcal{O}'[[\mathcal{G}]] \xrightarrow{j} \mathcal{O}'(1) \to 0$$

is exact (notice that $i$ now maps $\mathcal{U}$ into $\mathcal{O}'[[\mathcal{G}]]$). Exactness of (13) will follow from the (topological) flatness of $\mathcal{O}_K$ over $\mathcal{O}'$. There are two points to check. Injectively of $i$ (this is a-priori clear only when $k = k'$), and that any $\mu \in Ker(j)$ is $\mu_\beta$ for a suitable $\beta$. We begin with a series of three lemmas due to Coleman [Col2].

**Lemma.** *Let $\mathcal{N}$ be the norm operator for $\hat{\mathbf{G}}_m$ over $\mathcal{O}'$ (see 2.1). Let $\mathcal{O}'/\wp' = \mathbf{F}_q$. Then for any $\overline{g} \in \mathbf{F}_q[[S]]^x$ there exists $g \in \mathcal{O}'[[S]]^x$, $\mathcal{N}g = g^\varphi$ and $\overline{g} = g \bmod \wp'$.*

PROOF: Let $g_0$ be an arbitrary lifting of $\overline{g}$. Let $g_i = \varphi^{-i}\mathcal{N}^{(i)}g_0$. It follows from 2.1(i) and (iv) that $g = \lim g_i$ exists and satisfies $\mathcal{N}g = g^\varphi$, $g \equiv g_0 \bmod \wp'$ (compare the proof of theorem 2.2).

**3.11 Lemma.** *Consider the map $\partial : \mathbf{F}_q[[X]]^x \to \mathbf{F}_q[[X]]$, $\partial g = X \cdot g'/g$. The image of $\partial$ consists of all the power series $h = \sum_{n=1}^{\infty} c_n X^n$ with $c_{pn} = c_n^p$.*

PROOF: Modulo $\mathbf{F}_q^x$, which belong to $Ker(\partial)$, $\mathbf{F}_q[[X]]^x$ is generated (topologically) by $1 - aX^k$ for $a \in \mathbf{F}_q^x$ and $k \geq 1$. Since

$$(21) \qquad \partial(1 - aX^k) = -k \cdot \sum_{j=1}^{\infty} a^j X^{jk}$$

satisfies the condition $c_{pn} = c_n^p$, it remains to show that this is the *only* restriction. Let $h = \sum_{n=1}^{\infty} c_n X^n$ be as above. If $p \nmid k$, then $\partial(1 - aX^k) \equiv -kaX^k \bmod X^{k+1}$, so adding up successively power series of the form (21), for $k$ relatively prime to $p$, we find in the image of $\partial$ a power series that agrees with $h$ at least at $c_n$ for $p \nmid n$. But since $c_{pn} = c_n^p$ holds for $h$, as well as for (21), this power series automatically coincides with $h$.

**3.12 Lemma.** *Let $\mathcal{E}$ be the subgroup of all $h \in \mathcal{O}'[[X]]$ satisfying*

$$h^{\varphi}((1+X)^p - 1) = \frac{1}{p} \sum_{\varsigma^p=1} h(\varsigma(1+X) - 1). \tag{22}$$

*Then $X(1+X)^{-1} \cdot \mathcal{E} \bmod \wp' = Image(\partial)$.*

PROOF: For any $h$ the right side of (22) is an integral power series, and an argument parallel to 2.1 shows that there exists a unique $\mathcal{S}h \in \mathcal{O}'[[X]]$ such that

$$\mathcal{S}h((1+X)^p - 1) = \frac{1}{p} \sum_{\varsigma^p=1} h(\varsigma(1+X) - 1). \tag{23}$$

We are interested in $\mathcal{E} = \{h \mid \mathcal{S}h = h^{\varphi}\}$. Let $\delta : \mathcal{O}'[[X]]^x \to \mathcal{O}'[[X]]$, $\delta g = (1+X)g'/g = D \log g$. ($D = (1+X)\dfrac{d}{dX}$ is the translation invariant derivation on $\hat{\mathbf{G}}_m$). Thus $\partial \equiv X(1+X)^{-1}\delta \bmod \wp'$. If $\mathcal{N}g = g^{\varphi}$ then $\mathcal{S}(\delta g) = (\delta g)^{\varphi}$ so in view of lemma 3.10 $X(1+X)^{-1} \cdot \mathcal{E} \bmod \wp' \supset Image(\partial)$. Suppose this was a strict inclusion. From the description of $Image(\partial)$ given in 3.11, would then follow that there is some $h \in \mathcal{E}$ with $X(1+X)^{-1}h \equiv \sum_{n=1}^{\infty} c_n X^{pn} \bmod \wp'$, and not all $c_n \equiv 0 \bmod \wp'$. Equivalently, $h \equiv \sum_{n=1}^{\infty} c_n(X^{pn-1} + X^{pn}) \bmod \wp'$. However, $\mathcal{S}(X^{pn}) \equiv X^n$ and $\mathcal{S}(X^{pn-1}) \equiv X^{n-1} \bmod \wp'$, as a simple calculation reveals. This contradicts $\mathcal{S}h = h^{\varphi}$, and the lemma is proved.

**Corollary.** *The map $\delta g = (1+X)g'/g$ maps $\{g \in \mathcal{O}'[[X]]^x \mid \mathcal{N}g = g^{\varphi}\}$ onto $\{h \in \mathcal{O}'[[X]] \mid \mathcal{S}h = h^{\varphi}\}$. Its kernel consists of the roots of unity in $\mathcal{O}'$.*

PROOF: Let $\mathcal{M}$ be the first group and $\mathcal{E}$ the second. Clearly $\delta$ maps $\mathcal{M}$ into $\mathcal{E}$, and the lemma shows $\delta(\mathcal{M}) \bmod \wp' = \mathcal{E}/p\mathcal{E}$, so $\delta(\mathcal{M}) = \mathcal{E}$. The statement about the kernel is obvious. The corollary is all we shall need to finish the proof.

**3.13 Lemma.** *Let $\nu$ be an $\mathcal{O}'$ – valued measure on $\mathbf{Z}_p^x$ and assume that $Tr_{k'/k}(\nu(\mathbf{Z}_p^x)) = 0$. Then there exists $h \in \mathcal{O}'[[X]]$ satisfying (22) (i.e. $\mathcal{S}h = h^{\varphi}$) such that, with*

$$\tilde{h}(X) = h(X) - \frac{1}{p}\sum_{\varsigma^p=1} h(\varsigma(1+X) - 1) = h(X) - h^{\varphi}((1+X)^p - 1)$$

(see (7')), $\tilde{h}(X) = \int_{\mathbf{Z}_p^x} (1 + X)^\alpha d\nu(\alpha)$.

PROOF: Extend $\nu$ to a measure on $\mathbf{Z}_p$ by (1) $\nu(p^i U) = \varphi^i \nu(U)$ $(U \subset \mathbf{Z}_p^x)$, (2) $\nu(p^{id}\mathbf{Z}_p) = 0$ $(d = [k' : k])$. These two conditions are compatible because $Tr_{k'/k}(\nu(\mathbf{Z}_p^x)) = 0$. Let $h = P_\nu$, i.e.

$$h(X) = \int_{\mathbf{Z}_p} (1 + X)^\alpha d\nu(\alpha).$$

Then $h^\varphi((1 + X)^p - 1) = \int_{\mathbf{Z}_p} (1 + X)^{p\alpha} d\nu^\varphi(\alpha) = \int_{p\mathbf{Z}_p} (1 + X)^\alpha d\nu(\alpha) = h(X) - \tilde{h}(X)$, so $h^\varphi = \mathcal{S}h$ and the lemma is proved.

**3.14** We are now in a position to conclude the proof of (20). Recall our notation: $[k' : k] = d$, $k_p = \cup k(\varsigma_n)$, $\xi = p^d$, $k_\xi = k_p k'$, and $G = Gal(k_\xi/k')$. Let $H = Gal(k_\xi/k_p)$ and denote by $\varphi$ the unique element of $H$ inducing $\varphi$ on $k'$. Then $\mathcal{G} = G \times H = G \times <\varphi>$. Note that in the notation of 3.7 (14) and (15), $\kappa(\varphi) = 1$, and $g_{\varphi(\beta)} = g_\beta^\varphi$. Let $\mathcal{L}$ be the subgroup of $\mathcal{O}'[[\mathcal{G}]]$ consisting of all $\mu$ satisfying

$$(a)\ j(\mu) = 0 \qquad (b)\ \mu(\varphi^{-1}U) = \varphi(\mu(U)).$$

Since $k'/k$ is unramified, it is easily seen that $Ker(j) = \mathcal{O}' \otimes_{\mathbf{Z}_p} \mathcal{L}$. It is therefore enough to prove that $i$ maps $\mathcal{U}$ isomorphically onto $\mathcal{L}$. That $i(\mathcal{U}) \subset \mathcal{L}$ follows at once from $g_{\varphi(\beta)} = g_\beta^\varphi$. Let $\mu \in \mathcal{L}$, and view its restriction to $G$ as a measure on $\mathbf{Z}_p^x$ through the cyclotomic character $\kappa$ $(= \mathbf{N})$. Because of $(b)$, condition $(a)$ reads $Tr_{k'/k}(\int_{\mathbf{Z}_p^x} \alpha d\mu(\alpha)) = 0$. Let $d\nu(\alpha) = \alpha d\mu(\alpha)$, and let $h$ be the power series of lemma 3.13. Let $g$ be as in corollary 3.12, normalized (dividing by a root of unity from $\mathcal{O}'$) so that $g(0) \equiv 1 \ mod\ \wp'$. Then $\beta_n = (\varphi^{-n}g)(\varsigma_n - 1) \in U(k_\xi^n)$. $\mathcal{N}g = g^\varphi$ implies $N_{m,n}\beta_m = \beta_n$, so $\beta \in \mathcal{U}$, $g = g_\beta$, $h = D \log g_\beta$, and

$$\tilde{h} = D\, \widetilde{\log g_\beta} = \int_{\mathbf{Z}_p^x} (1 + X)^\alpha \alpha d\mu(\alpha).$$

It follows that $\alpha \cdot d\mu_\beta(\alpha) = \alpha \cdot d\mu(\alpha)$, which proves $\mu = \mu_\beta$ on $G$. Comparing $(b)$ with the way $\mu_\beta$ was extended to $\mathcal{G}$ (3.4), $\mu = \mu_\beta = i(\beta)$. Therefore $i(\mathcal{U}) = \mathcal{L}$, and the proof of theorem 3.7 is complete.

REMARKS: It was not necessary to reduce to the case of $\hat{\mathbf{G}}_m$. The analysis of 3.10-3.14 could be carried out with the original $F_f$, *mutatis mutandis.* However, proposition 3.9 is independently interesting, so we chose to prove it first.

Iwasawa (and later Wintenberger [Win] in a more general context) determined the $\Lambda$-structure of $\mathcal{U}$ as a $\mathbf{Z}_p[[\mathcal{G}]]$ module ([Iw] 12.2). The significance of our theorem is that $\mathcal{U} \hat{\otimes} \mathcal{O}_K$ is *canonically* embedded in $\mathcal{O}_K[[\mathcal{G}]]$ with the right cokernel.

Theorem 3.7 will be used later on (III.1). Special cases of it are equivalent to lemmas 23-26 and theorem 27 of [Ya], but we found the proofs there somewhat ad-hoc. In particular, they rely on the general structure theorems of Wintenberger, while the approach taken here does not. In fact, in our case we can deduce Wintenberger's results from theorem 3.7.

## 4. THE EXPLICIT RECIPROCITY LAW

Our aim is to prove a version of the explicit reciprocity law in local class field theory, that was discovered by Wiles [Wi], generalizing earlier work of Artin-Hasse [A-H] and Iwasawa [Iw4]. For further generalizations, as well as discussion of the literature, see [dS2] and the bibliography there. This section will be needed only in chapter IV, so the reader may skip it and proceed directly to the next chapter.

Notation and assumptions are as in §§1,2.

**4.1** KUMMER THEORY ON LUBIN TATE GROUPS: Let $F_f$ be a relative Lubin Tate group defined over $\mathcal{O}'$, and $k_\xi^n = k'(W_f^n)$ the field of division points of level $n$. Let $\alpha_n \in \wp_n$ (the valuation ideal of $k_\xi^n$). Recall that $f^{(n)} = \varphi^{n-1} f \circ \ldots \circ \varphi f \circ f \in Hom(F_f, F_{\varphi^n f})$. Let $a_n$ be any root (in the open unit disk of $\overline{k}$) of

$$(1) \qquad (\varphi^{-n} f^{(n)})(a_n) = \alpha_n.$$

The extension $k_\xi^n(a_n) = L$ of $k_\xi^n$ is *abelian.* Indeed, $Gal(L/k_\xi^n)$ is embedded in $W_{\varphi^{-n} f}^n$ via the map $\sigma \mapsto \sigma(a_n)[-]a_n$. This map is a group homomorphism independent of $a_n$ because $W_{\varphi^{-n} f}^n \subseteq k_\xi^n$. The *Kummer pairing* $F_f(\wp_n) \times Gal(k_\xi^{n,ab}/k_\xi^n) \rightarrow$

$W^n_{\varphi^{-n}f}$ is the pairing

$$(2) \qquad < \alpha_n, \sigma > = \sigma(a_n)[-]a_n$$

($L^{ab}$, for any $L$, is its maximal abelian extension). It is bilinear, because in addition to the linearity in $\sigma$ mentioned above, $< \alpha_n[+]\alpha'_n, \sigma > = < \alpha_n, \sigma > [+] < \alpha'_n, \sigma >$. The pairing is even $\mathcal{O}$-linear in the first variable: $< [e]\alpha_n, \sigma > = [e] < \alpha_n, \sigma >$ for any $e \in \mathcal{O}$. Clearly the kernel on the left is $f^{(n)}(\wp_n)$, but it is not at all clear what the kernel on the right is, i.e., what is the field obtained by "extracting $f^{(n)}$ roots" of all $\alpha \in \wp_n$. We only know so far that it is an abelian extension of $k^n_\xi$ of exponent $q^n$ (however, see [dS2], theorem 2).

Using the local Artin map, we define a pairing $F_f(\wp_n) \times (k^n_\xi)^x \to W^n_{\varphi^{-n}f}$ by

$$(3) \qquad (\alpha_n, \beta_n) = (\alpha_n, \beta_n)_{n,f} = < \alpha_n, \sigma_{\beta_n} >$$

where $\sigma_\beta$ is the Artin symbol of $\beta$. Our aim is to compute (3) by means of some analytic formula.

**4.2** For simplicity of notation and exposition we shall assume from now on that $k' = k$, i.e. $F_f$ is an *absolute* Lubin Tate group. The "relative" case is treated similarly, but because of the need to work with all $F_{\varphi^{-i}f}$ at once, the notation becomes cumbersome.

Assume, therefore, that $f(X) = \pi X + \cdots$, $\pi$ a prime of $k$, and fix a generator of the Tate module $(\omega_n)$ as in 2.2, $\omega_n \in \tilde{W}^n_f$, $[\pi](\omega_n) = \omega_{n-1}$.

**Theorem** (Wiles [Wi]). *Let $\beta_n \in (k^n_\pi)^x$ satisfy $N_n(\beta_n) \in < \pi >$ ($N_n$ is the norm from $k^n_\pi$ to $k$; this assumption means that $\beta_n$ sits in a norm-coherent sequence $\beta \in B = \varprojlim (k^n_\pi)^x$). Choose a norm coherent sequence $\beta$ with $\beta_n$ as its $n^{th}$ coordinate, and let $g_\beta$ be the Coleman power series of $\beta$ (2.2). Let $\lambda(X)$ be a logarithm of $F_f$, so that $\lambda'(X) \in \mathcal{O}[[X]]^x$. Put $\delta g = \frac{1}{\lambda'} \cdot \frac{g'}{g} \in X^{-1} \cdot \mathcal{O}[[X]]$. Let*

$$(4) \qquad [\alpha_n, \beta_n] = [\alpha_n, \beta_n]_{n,f} = [\pi^{-n} Tr_{k^n_\pi/k}(\lambda(\alpha_n)\delta g_\beta(\omega_n))]_f(\omega_n).$$

*(The fact that the quantity in square brackets is in $\mathcal{O}$, and modulo $\pi^n$ depends only on $\alpha_n, \beta_n$, is part of the theorem). Then*

$$(5) \qquad (\alpha_n, \beta_n)_{n,f} = [\alpha_n, \beta_n]_{n,f}.$$

The *analytical pairing* (4) is (in principle and in practice) computable. Indeed, one only needs to compute a certain element of $\mathcal{O}/\pi^n$, and this requires a *finite* amount of computation.

Our proof is different from Wiles' in one important way: It is based on the observation that it is easier to prove (5) for all Lubin-Tate groups at once, adjusting $F_f$ "to fit" $\beta$ (see the "reduction lemma" below).

**4.3** Let $B = \varprojlim (k_\pi^n)^x$, the inverse limit taken with respect to the norm maps. Thus $B$ is an extension of $\mathbf{Z}$ by the profinite group $\mathbf{F}_q^x \times \mathcal{U}$, where $\mathcal{U}$ is the inverse system of principal units.

Let $A_f = \varinjlim F_f(\wp_n)$, where $F_f(\wp_n)$ maps to $F_f(\wp_{n+1})$ by $f = [\pi]$.

The first observation is that the Kummer pairings (3) combine to give a single pairing

$$(6) \qquad (\ ,\ )_f : A_f \times B \to W_f.$$

To check this, let $m \geq n$, $\alpha_m = f^{m-n}(\alpha_n)$, $\beta_n = N_{m,n}\beta_n$, and observe that $(\alpha_n, \beta_n)_n = (\alpha_m, \beta_m)_m$. Similarly, denoting the trace from $k_\pi^n$ to $k$ by $Tr_n$, $Tr_m(\lambda(\alpha_m)\delta g_\beta(\omega_m)) = Tr_n(\pi^{m-n}\lambda(\alpha_n)Tr_{m,n}\delta g_\beta(\omega_m)) = Tr_n(\pi^{m-n}\lambda(\alpha_n)\pi^{m-n}\delta g_\beta(\omega_n))$ because property 2.1(iii), asserting $g_\beta \circ f^i = \prod_{\omega \in W_f^i} g_\beta(X[+]\omega)$, gives, upon applying $D \log = \delta$ to it,

$$(7) \qquad \pi^i \delta g_\beta \circ f^i(X) = \sum_{\omega \in W_f^i} \delta g_\beta(X[+]\omega).$$

It follows that $[\alpha_n, \beta_n]_{n,f} = [\alpha_m, \beta_m]_{m,f}$, hence the symbol $[\ ,\ ]_f$ is also defined on $A_f \times B$ (provided it makes sense!).

The second observation is that both pairings $(\ ,\ )_f$ and $[\ ,\ ]_f$ are bilinear, so it is enough to prove (5) for *prime* $\beta$ (i.e. $\beta_n$ is prime for every $n$), as these generate $B$.

The third observation is that if—with a *given* $\beta \in B$—(5) holds for *one* $f \in \mathcal{F}_\pi$ and every $\alpha \in A_f$, then it holds—with the same $\beta$—for *all* $f \in \mathcal{F}_\pi$ and $\alpha \in A_f$. Indeed, if $\eta : F_f \simeq F_{f'}$ is an isomorphism over $\mathcal{O}$ (where $f, f' \in \mathcal{F}_\pi$), $\alpha \in A_f$ and $\beta \in B$, then $\eta((\alpha,\beta)_f) = (\eta(\alpha),\beta)_{f'}$ and also $\eta([\alpha,\beta]_f) = [\eta(\alpha),\beta]_{f'}$.

Now, given $\beta$ for which $\beta_n$ is a prime of $k_\pi^n$, $g_\beta \in X\mathcal{O}[[X]]^x$. Let $g_\beta^{-1}$ denote the inverse of $g_\beta$ with respect to composition of power series. Then $f' = g_\beta \circ f \circ g_\beta^{-1} \in \mathcal{F}_\pi$ too, and $g_\beta : F_f \simeq F_{f'}$. Furthermore $g_\beta(\omega_n) = \beta_n$ so $(\beta_n)$ is a generator of the Tate module of $F_{f'}$. By the last remark, we may prove (5) for $f'$ instead of $f$. In other words, without loss of generality we may assume $\omega_n = \beta_n$.

REDUCTION LEMMA. *To prove theorem 4.2 we may assume that $N_{m,n}\omega_m = \omega_n$, and that $\beta_n = \omega_n$ for each $n$. The formula to be proved is then*

$$(\alpha_n, \omega_n)_n = \left[\pi^{-n} Tr_n \left(\frac{\lambda(\alpha_n)}{\omega_n \cdot \lambda'(\omega_n)}\right)\right]_f (\omega_n), \tag{8}$$

*and its validity should only be established for high enough $n$.*

PROOF: Everything follows from the discussion above, except for formula (8), but with $\beta = \omega$, $g_\beta = X$, so $\delta g_\beta = \dfrac{1}{X \cdot \lambda'(X)}$.

Note that in general $\omega_n$ is not a norm coherent sequence. It is easy to check that this is the case if and only if $\mathcal{N}X = X$.

**4.4** PROOF OF (8): Throughout the proof, it will be convenient to use $\equiv$ *mod* $\pi^{an+c}$ to mean "there exists an integer $c$, *independent* of $n$, such that the congruence holds for large enough $n$". The letter $c$ will stand for different constants in different places.

STEP 1: If $\alpha \in \wp_n$, then $(\alpha,\alpha)_n = 0$.

PROOF: As noted above, $\mathcal{N}X = X$, so if $[\pi^n](a) = \alpha$, $\alpha = \prod_{\omega \in W_f^n} (a[+]\omega)$. Let $L = k_\pi^n(a)$, and let $V \subset W_f^n$ be the subgroup isomorphic to $Gal(L/k_\pi^n)$

under the map (2). Let $R$ be a system of representatives for $W_f^n$ modulo $V$. Then $\alpha = \prod_{u\in R}\prod_{v\in V}(a[+]u[+]v) = \prod_{u\in R} N_{L/k_\pi^n}(a[+]u)$. By class field theory, the Artin symbol of $\alpha$ on $L$ is trivial.

STEP 2: Fix $\alpha \in A_f$. For large $n$, $\alpha_n[+]\omega_n$ is a prime of $k_\pi^n$, because $\alpha_n \to 0$ rapidly (recall $[\pi](\alpha_n) = \alpha_{n+1}$ so $\nu(\alpha_{n+1}) \geq min(q\nu(\alpha_n), \nu(\alpha_n) + \nu(\pi)))$. Write

$$(9) \qquad \alpha_n[+]\omega_n = \omega_n \cdot (1 + \gamma_n).$$

By step 1, and bilinearity of $(\ ,\ )_n$,

$$(10)\ 0 = (\alpha_n[+]\omega_n, \omega_n(1+\gamma_n)) = (\alpha_n,\omega_n)[+](\alpha_n, 1+\gamma_n)[+](\omega_n, 1+\gamma_n).$$

STEP 3: For large $n$, $(\alpha_n, 1+\gamma_n)_n = 0$. Indeed, if $m \geq n$, $(\alpha_m, 1+\gamma_m)_m = (\alpha_n, N_{m,n}(1+\gamma_m))_n$. When $m \to \infty$, $\alpha_m \to 0$, so $\gamma_m \to 0$ and $N_{m,n}(1+\gamma_m)$ tends to 1. By continuity of the Artin symbol, $(\alpha_m, 1+\gamma_m)_m$ tends to 0. But with fixed $n$, $(\alpha_m, 1+\gamma_m)_m$ is always in $W_f^n$, which is *discrete*. Hence for large $m$, $(\alpha_m, 1+\gamma_m)_m = 0$.

STEP 4: For large $n$

$$(11) \qquad \begin{aligned}(\alpha_n,\omega_n)_n &= [-](\omega_n, 1+\gamma_n)_n \quad \text{(steps 2 and 3)}\\ &= \omega_{2n}[-][N_n(1+\gamma_n)^{-1}]_f(\omega_{2n})\\ &= [1 - N_n(1+\gamma_n)^{-1}](\omega_{2n}).\end{aligned}$$

The second equality shows the advantage of dealing with $\omega_n$. Its "$\pi^n$-root" $\omega_{2n}$ generates a field $L = k_\pi^{2n}$ *abelian over* $k$, so for the Artin map $(1+\gamma_n, L/k_\pi^n) = (N_n(1+\gamma_n), L/k)$. The effect of this symbol on $\omega_{2n}$ is known by the theory of Lubin and Tate!

STEP 5: $1 - N_n(1+\gamma_n)^{-1} \equiv Tr_n(\gamma_n)\ mod\ \pi^{3n-c}$.

PROOF: Note that $\alpha_n \equiv 0\ mod\ \pi^{n-c}$, hence the same congruence holds for $\gamma_n$ (cf. (9)). Since the different of $k_\pi^n/k$ is "of size" $\pi^{n-c}$ [Se3, IV §1], $Tr_n(\gamma_n) \equiv$

$0 \; mod \; \pi^{2n-c}$, and $Tr_n(\gamma_n^2) \equiv 0 \; mod \; \pi^{3n-c}$. Next

$$
\begin{aligned}
1 - N_n(1 + \gamma_n)^{-1} &\equiv 1 - \prod_{\sigma} (1 - \gamma_n + \gamma_n^2)^{\sigma} \; mod \; \pi^{3n-c} \\
&\equiv \sum \gamma_n^{\sigma} - \frac{1}{2} \sum (\gamma_n^2)^{\sigma} - \frac{1}{2} \sum \sum \gamma_n^{\sigma_1} \gamma_n^{\sigma_2} \qquad (12) \\
&\equiv Tr_n(\gamma_n) - \frac{1}{2} Tr_n(\gamma_n^2) - \frac{1}{2}(Tr_n(\gamma_n))^2 \\
&\equiv Tr_n(\gamma_n) \; mod \; \pi^{3n-c}.
\end{aligned}
$$

STEP 6: Take $\lambda$ of both sides of (9) and Taylor-expand in $\gamma_n$. We find

$$\lambda(\alpha_n) \equiv \gamma_n \cdot \omega_n \cdot \lambda'(\omega_n) \; mod \; \pi^{2n-c},$$

hence

$$\gamma_n \equiv \frac{\lambda(\alpha_n)}{\omega_n \cdot \lambda'(\omega_n)} \; mod \; \pi^{2n-c}.$$

Thus $Tr_n(\gamma_n) \equiv Tr_n\left(\frac{\lambda(\alpha_n)}{\omega_n \cdot \lambda'(\omega_n)}\right) \; mod \; \pi^{3n-c}$.

STEP 7: Combining steps 4, 5, and 6 we get

$$(\alpha_n, \omega_n)_n = \left[Tr_n\left(\frac{\lambda(\alpha_n)}{\omega_n \cdot \lambda'(\omega_n)}\right)\right]_f (\omega_{2n})$$

(for large $n$). Since the left hand side lies in $W_f^n$, the quantity in square brackets is divisible by $\pi^n$, and this concludes the proof of (8).

# CHAPTER II
# p-ADIC L FUNCTIONS

The method of chapter I, 3.1-3.6 will be applied here to construct $p$-adic $L$ functions over quadratic imaginary fields. These $p$-adic $L$ functions appeared first in a paper of Manin and Višik from 1974 [M-V], and almost simultaneously Katz gave another construction [K1]. Coates and Wiles [C-W2] introduced the point of view taken here when they showed how to produce the "one variable" functions (see below) from norm-coherent sequences of elliptic units. They related the $p$-adic $L$ functions to class field theory, and formulated a "main conjecture". Their approach was developed further mainly by R. Yager [Ya1], [Ya2], and also by P. Cassou-Nogues [CN], R. Gillard [Gi1], J. Tilouine [Ti] and the author, hand in hand with the emergence of Coleman's power series as a powerful tool. Our aim here is to give a general and self contained exposition, free from any restrictive assumptions.

The extension of these results to arbitrary CM fields poses some tantalizing problems. Out of the three approaches (Manin-Višik, Katz, and Coates-Wiles), only Katz' method of $p$-adic interpolation of real analytic Eisenstein series seems to work [K2]. The approach taken in this book, for example, would require a supply of special units in abelian extensions of the ground field, similar to the elliptic units. Exhibiting such units is a major open problem in number theory. Moreover, a consequence of a Coates-Wiles type construction (when compared with [K2]) would be a deep theorem on monomial relations between $p$-adic periods of abelian varieties with complex multiplication, analogous to Shimura's period relations between the corresponding complex periods. We have managed to demonstrate these $p$-adic period relations in some cases [dS3].

The $p$-adic $L$ functions constructed in this chapter belong to a long array of functions bearing this title. Among the *abelian* ones, there are first and foremost those of Kubota and Leopoldt [K-Le], and their generalizations by Deligne and

Ribet [D-R], as well as the ones mentioned above. Then there are "non-abelian" $p$-adic $L$ functions arising from modular forms. Of fundamental importance here are the works of Mazur and Swinnerton-Dyer [M-SD], and of Manin [Man], but they all lie outside the scope of this book. See also the paper by Haran [Har].

An outline of chapter II follows. In §1 we summarize various results from the theory of elliptic curves with complex multiplication, needed later on. Section 2 introduces special global units, the elliptic units of Siegel, Ramachandra and Robert. They come in norm-coherent sequences, and when viewed inside the local units in the right tower of fields, the procedure of I.3.1-3.6 may be applied to get measures out of them. This simple idea translates into elaborate computations in §4. The measures so obtained *are* the $p$-adic $L$ functions, up to simple twisting factors (see the discussion of measures and $L$-functions in I.3). The relations that exist between elliptic units, special values of Eisenstein series, and $L$-series, worked out in §3, allow us to conclude that the $p$-adic $L$ functions indeed interpolate special values of classical Hecke $L$ series, so they deserve their name. The rest of chapter II is devoted to the exploration of some *analytical* properties of these $p$-adic $L$ functions. Algebraic and arithmetic aspects are left out for the next chapters.

## 1. Background

**1.1** GROSSENCHARACTERS: Throughout this chapter, $K$ denotes a quadratic imaginary field of discriminant $-d_K$ ($d_K$ is a positive integer congruent to 0 or 3 *modulo* 4). The number of roots of unity in $K$, $w_K$, is 2, 4 or 6. Fractional ideals of $K$ are denoted by gothic letters $\mathfrak{a}$, $\mathfrak{f}$, $\mathfrak{p}$ etc. If $\mathfrak{f}$ is an integral ideal let $w_\mathfrak{f}$ be the number of roots of unity congruent to 1 *mod* $\mathfrak{f}$. The *ray class field modulo* $\mathfrak{f}$ is denoted $K(\mathfrak{f})$, and $K(\mathfrak{f}\mathfrak{g}^\infty)$ is the union of $K(\mathfrak{f}\mathfrak{g}^n)$ for all $n$. $K(1)$ is the Hilbert class field, and $[K(\mathfrak{f}) : K(1)] = |(\mathcal{O}_K/\mathfrak{f})^x| \cdot w_\mathfrak{f}/w_K$.

We view all our number fields as subfields of a fixed algebraic closure $\overline{\mathbf{Q}}$ of the rationals, and choose, once and for all, a rational prime $p$ and embeddings $i_\infty$ and

$i_p$ of $\overline{\mathbf{Q}}$ in $\mathbf{C}$ and $\mathbf{C}_p$. When there is no danger of confusion, we shall drop $i_\infty$ and $i_p$ from the notation.

Let $F$ be a number field containing $K$, and $\chi$ a *grossencharacter of type* $A_0$ of $F$. Thus $\chi$ is a homomorphism from the group of all fractional ideals relatively prime to some $\mathfrak{f}$ into $\overline{\mathbf{Q}}^x$. There exist an integral ideal $\mathfrak{f}$ and an element $\omega = \sum n(\sigma)\sigma \in \mathbf{Z}[Emb(F, \overline{\mathbf{Q}})]$ such that $\chi((a)) = a^\omega$ for any $a \in F^x$, $a \equiv 1 \ mod^x \ \mathfrak{f}$. The smallest $\mathfrak{f}$ with this property is called the *conductor* of $\chi$, denoted $\mathfrak{f}_\chi$, and $\omega$ is its *infinity type*. Put $\chi(\mathfrak{a}) = 0$ if $\mathfrak{a}$ is integral and not relatively prime to $\mathfrak{f}_\chi$. The *$L-$ series of* $\chi$ (*with modulus* $\mathfrak{m}$) is the complex analytic function $L_\mathfrak{m}(\chi, s) = \sum \chi(\mathfrak{a})\mathbf{N}\mathfrak{a}^{-s}$, where $\mathfrak{a}$ runs over all integral ideals relatively prime to $\mathfrak{m}$. When $\mathfrak{m} = (1)$, we simply write $L(\chi, s)$ and refer to it as a *primitive* $L$ function. Note that in our notation we do not insist that $\mathfrak{f}_\chi | \mathfrak{m}$. The analytic continuation and functional equation of $L(\chi, s)$ are well known. When $F = K$ we shall also say that $\chi$ has infinity type $(k, j)$ if $\chi((a)) = a^k \overline{a}^j$ for $a \equiv 1 \ mod^x \ \mathfrak{f}_\chi$. We then let

$$(1) \qquad \Gamma_\chi(s) = \frac{\Gamma(s - min(k,j))}{(2\pi)^{s-min(k,j)}},$$

$$(2) \qquad L_\infty(\chi, s) = \Gamma_\chi(s)L(\chi, s), \qquad R(\chi, s) = (d_K \mathbf{N}\mathfrak{f}_\chi)^{s/2} L_\infty(\chi, s).$$

With these definitions, the functional equation takes the form

$$(3) \qquad R(\chi, s) = W \cdot R(\overline{\chi}, 1 + k + j - s).$$

$W$ is a constant of absolute value 1, the *Artin root number.*

Following Deligne, call $\chi$ *critical* if the $\Gamma$ factors in the functional equation, namely $\Gamma_\chi(s)$ and $\Gamma_{\overline{\chi}}(1 + k + j - s)$, are finite at $s = 0$. This holds if and only if $k < 0$ and $0 \leq j$, or $0 \leq k$ and $j < 0$. Schematically, the critical infinity types $(k, j)$ are depicted as follows.

(4)

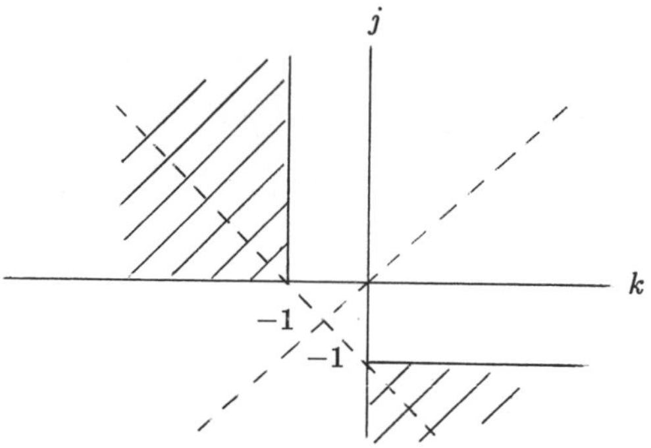


If the infinity type of $\chi$ corresponds to a point $P$, $\overline{\chi}$ and $\chi\mathbf{N}^{-k-j-1}$ are represented by the reflections of $P$ in the lines $k = j$ and $k + j = -1$ respectively. The integer $k + j$ is called the *weight* of $\chi$. For reasons that will become clear later, we call the dotted lines $k = j$ and $k + j = -1$ the *cyclotomic* and *anticyclotomic* lines respectively. Note that the cyclotomic line lies outside the critical range, while all the lattice points on the anticyclotomic line are within it.

Finally, let $\chi$ be a grossencharacter of $K$ of type $A_0$ and conductor $\mathfrak{f}$. It is well known ([We]) that the values of $\chi$ lie in a number field of finite degree, and that, when viewed $p$-adically via the embedding $i_p : \overline{\mathbf{Q}} \hookrightarrow \mathbf{C}_p$, $\chi$ extends continuously to a Galois character (also denoted $\chi$)

$$(5) \qquad \begin{cases} \chi : \mathcal{G} = Gal(K(\mathfrak{f}p^\infty)/K) \to \mathbf{C}_p^x \\ \chi(\sigma_\mathfrak{a}) = \chi(\mathfrak{a}), \qquad (\mathfrak{a}, \mathfrak{f}p) = 1, \qquad \sigma_\mathfrak{a} = (\mathfrak{a}, K(\mathfrak{f}p^\infty)/K). \end{cases}$$

If furthermore $p$ splits in $K$, $p = \mathfrak{p}\overline{\mathfrak{p}}$, and $\mathfrak{p}$ is the place induced by the inclusion $K \subset \overline{\mathbf{Q}} \hookrightarrow \mathbf{C}_p$, and if the infinity type of $\chi$ is $(k, 0)$ for some $k$, then the Galois character (5) factors through $Gal(K(\mathfrak{f}\mathfrak{p}^\infty)/K)$. If the infinity type is $(0, j)$, then (5) factors through $Gal(K(\mathfrak{f}\overline{\mathfrak{p}}^\infty)/K)$.

**1.2** ELLIPTIC CURVES: General references for elliptic curves are the book of Silverman [Sil] and the survey article by J. Tate [Ta]. The purpose of this section is to recall some terminology.

If $F$ is a number field and $E$ an elliptic curve over $F$, a *Weierstrass model* for $E/F$ is any plane model

$$(6) \quad y^2 = 4x^3 - g_2x - g_3, \qquad g_2, g_3 \in F, \qquad \Delta = g_2^3 - 27g_3^2 \neq 0.$$

Such a model is unique up to $(g_2, g_3) \mapsto (u^4g_2, u^6g_3)$, $u \in F^x$. The *standard differential* associated with (6) is the differential

$$(7) \qquad \omega_E = \frac{dx}{y}.$$

This is a basis for the (one dimensional) space of differentials of the first kind, rational over $F$.

The pair $(E, \omega_E)$ determines a *lattice* $L \subset \mathbf{C}$ *of periods*, $L = \{\int_\gamma \omega_E \mid \gamma \in H_1(E(\mathbf{C}), \mathbf{Z})\}$. Conversely, the lattice $L$ determines the Weierstrass model (6) via $g_2 = g_2(L)$, $g_3 = g_3(L)$ (we then write $\Delta = \Delta(L)$), where

$$(8) \qquad g_2(L) = 60 \cdot {\sum_{\omega \in L}}' \omega^{-4}, \qquad g_3(L) = 140 \cdot {\sum_{\omega \in L}}' \omega^{-6}.$$

The complex points of $E$ are uniformized by $\mathbf{C}/L$ through *Weierstrass' $\wp$-function* and its derivative

$$(9) \qquad \wp(z, L) = \frac{1}{z^2} + {\sum_{\omega \in L}}' \left\{ \frac{1}{(z-\omega)^2} - \frac{1}{\omega^2} \right\}.$$

The map $z \mapsto \xi(z, L) = (\wp(z, L), \wp'(z, L))$ is an analytic isomorphism of $\mathbf{C}/L$ onto $E(\mathbf{C})$.

The *$j$ – invariant* of $E$ is given by

$$(10) \qquad j_E = j(L) = 1728\, g_2^3/\Delta.$$

Two elliptic curves defined over $F$ are isomorphic over $\overline{F}$ if and only if they have the same $j$-invariant.

**1.3** COMPLEX MULTIPLICATION: General references are [Sh], [Gr], and the more classical expositions in [Bo] and [Deu]. The older monographs of Weber, Fricke and Fueter contain much valuable information not to be found in modern literature.

Let $E$ be an elliptic curve defined over a number field $F$. Then $E$ *admits complex multiplications* if $End(E)$, the ring of endomorphisms of $E$ as an algebraic group over $F$, is strictly larger than $\mathbf{Z}$. In such a case $End(E)$ is isomorphic to an order $\mathcal{O}$ in a quadratic imaginry field $K$, $F$ contains $K(\mathcal{O})$, the *ring-class-field* (ringklassenkorper) of $\mathcal{O}$, and $E$ is isomorphic over $\overline{F}$ to an elliptic curve defined over $K(\mathcal{O})$. We always identify $a \in \mathcal{O}$ with the endomorphism $\iota(a)$ whose differential is $a$, i.e. $\iota(a)^*\omega_E = a\omega_E$. Write $E[a]$ for the kernel of $\iota(a)$, and for any ideal $\mathfrak{a} \subset \mathcal{O}$, let $E[\mathfrak{a}] = \cap_{a\in\mathfrak{a}} E[a]$, $E[\mathfrak{a}^\infty] = \cup_{1\leq n} E[\mathfrak{a}^n]$, $E_{tor} = \cup_{\mathfrak{a}} E[\mathfrak{a}]$.

The *main theorem of complex multiplication* ([Sh] 5.3) asserts first that the extension $F(E[\mathfrak{a}])/F$ obtained by adjoining to $F$ the coordinates of points of $E[\mathfrak{a}]$ in one (hence any) projective model of $E$ over $F$, is abelian. Second, there exists a unique grossencharacter of type $A_0$ $\psi = \psi_{E/F}$ of $F$, with values in $K$, and with the following property. If $\mathfrak{A}$ is an ideal of $F$, relatively prime to the conductor of $\psi$, then $\psi(\mathfrak{A}) \in \mathcal{O}$, and

$$\sigma_{\mathfrak{A}}(u) = \iota(\psi(\mathfrak{A}))(u) \tag{11}$$

for any $u \in E[\mathfrak{c}]$, $(\mathfrak{c}, N_{F/K}\mathfrak{A}) = 1$. The infinity type of $\psi$ is the "half norm type" $\sum \sigma$, the sum extending over all the embeddings of $F$ which restrict to the identity on $K$. The *conductor* of $E$ over $F$ is by definition the conductor of $\psi$. This agrees with other, more general definitions of conductors (see [Se-Ta]).

**1.4** A SPECIAL CLASS OF ELLIPTIC CURVES: It has been known for almost a century that the arithmetic of abelian extensions of $K$ is related to elliptic curves with complex multiplication by $K$, in the same sense that the arithmetic of abelian fields is related to the multiplicative group. We shall now introduce a special class of curves, particularly well-suited for that purpose.

From now on, until the end of §1, fix an *abelian* extension $F$ of $K$, of conductor

$\mathfrak{f}_{F/K}$, and let $E$ be an elliptic curve over $F$ satisfying

$$(12) \qquad \begin{cases} (i) & E \text{ has complex multiplications by } \mathcal{O}_K, \\ (ii) & F(E_{tor}) \text{ is an abelian extension of } K. \end{cases}$$

It follows from (i) that $F \supset K(1)$. To understand (ii), recall the theorem ([Sh] 7.44) saying that it is equivalent to the existence of a grossencharacter $\varphi$ of $K$, of type $(1,0)$, satisfying

$$(13) \qquad \psi_{E/F} = \varphi \circ N_{F/K}.$$

Fix once and for all such a $\varphi$. The other candidates are $\varphi\chi$, $\chi \in \widehat{Gal(F/K)}$. Since $\mathfrak{f}_{F/K} = \ell.c.m. \{\mathfrak{f}_\chi\}$,

$$(14) \qquad \mathfrak{f} = \ell.c.m. (\mathfrak{f}_\varphi, \mathfrak{f}_{F/K})$$

is equal to $\ell.c.m. \{\mathfrak{f}_{\varphi\chi}\}$, and therefore depends only on $F$ and $\psi$. Note that $\mathfrak{f}_\varphi$ is a proper ideal.

Another consequence of (ii) is that for any $\sigma \in Gal(F/K)$, $\psi_{E^\sigma/F} = \psi_{E/F}$. Since the grossencharacter of $E/F$ completely determines its $F$-isogeny class ([Gr] 9.2), all the Galois conjugates of $E$ are $F$-isogenous.

**Lemma.** *(i) Let $\varphi$ be a grossencharacter of $K$ of type $(1,0)$, and $F = K(\mathfrak{f}_\varphi)$. Let $j$ be the $j-$invariant of an elliptic curve with complex multiplication by $\mathcal{O}_K$. Then there exists a unique elliptic curve $E$ over $F$ with $j_E = j$, satisfying (12), for which $\psi_{E/F} = \varphi \circ N_{F/K}$.*

*(ii) Let $\mathfrak{f}$ be an integral ideal of $K$. Then there exists a grossencharacter $\varphi$ of type $(w_\mathfrak{f}, 0)$ and conductor $\mathfrak{f}$.*

PROOF: (i) Let $\psi = \varphi \circ N_{F/K}$. If $\mathfrak{A}$ is an ideal of $F$ relatively prime to $\mathfrak{f}_\varphi$, then $N_{F/K}(\mathfrak{A}) = (a)$ for some $a \equiv 1 \; mod^\times \; \mathfrak{f}_\varphi$, so $\psi(\mathfrak{A}) = a$. Let $E_0$ be any elliptic curve defined over $F$ with CM by $\mathcal{O}_K$ and $j$-invariant $j$. Let $\psi_0$ be its grossencharacter. Then $\varepsilon = \psi/\psi_0$ is a character of finite order with values in $\mathcal{O}_K^\times$. Let $E = E_0^\varepsilon$ be the *twist* of $E_0$ by $\varepsilon$ ([Gr] 3.3). Then ([Gr] 9.2) $\psi_{E/F} = \psi$, and since $E$ and $E_0$ are isomorphic over $\overline{\mathbf{Q}}$, $j_E = j$. The uniqueness is a consequence of the fact that the isomorphism class over $\overline{\mathbf{Q}}$ and the isogeny class over $F$ together determine the isomorphism class over $F$ ([Gr] §9).

(ii) This is obvious, and left to the reader.

**1.5** Notation and assumption as above, let $\mathfrak{a}$ be an integral ideal relatively prime to $\mathfrak{f}$. By the *main theorem on complex multiplication* ([Sh] 5.3) we deduce the following strengthening of (11).

**Proposition.** *There exists a unique isogeny* $\lambda(\mathfrak{a}) : E \to E^{\sigma_\mathfrak{a}}$ *over* $F$, *of degree* $\mathbf{N}\mathfrak{a}$, *characterized by*

$$\sigma_\mathfrak{a}(u) = \lambda(\mathfrak{a})(u) \tag{15}$$

*for any* $u \in E[\mathfrak{c}]$, $(\mathfrak{c}, \mathfrak{a}) = 1$.

Let $\omega$ be an $F$-rational differential of the first kind on $E$. Define the quantity $\Lambda(\mathfrak{a}) \in F$ by

$$\omega^{\sigma_\mathfrak{a}} \circ \lambda(\mathfrak{a}) = \Lambda(\mathfrak{a})\omega. \tag{16}$$

It is easily verified that $\Lambda(\cdot)$ satisfies the *cocycle condition* (cf. I.3.7 (14))

$$\Lambda(\mathfrak{a}\mathfrak{b}) = \Lambda(\mathfrak{a})^{\sigma_\mathfrak{b}}\Lambda(\mathfrak{b}) = \Lambda(\mathfrak{b})^{\sigma_\mathfrak{a}}\Lambda(\mathfrak{a}). \tag{17}$$

This enables us to extend the definition of $\Lambda$ to any fractional ideal relatively prime to $\mathfrak{f}$ so that (17) remains valid. A different choice of $\omega$ results in modifying $\Lambda$ by a coboundary. When a Weierstrass model (6) is given, we tacitly assume that $\Lambda(\cdot)$ is associated with the standard differential $\omega_E$.

If $(\mathfrak{a}, F/K) = 1$, $E^{\sigma_\mathfrak{a}} = E$, so $\Lambda(\mathfrak{a}) \in K^x$ and $\lambda(\mathfrak{a}) = \iota(\Lambda(\mathfrak{a}))$. Moreover

$$\Lambda(\mathfrak{a}) = \varphi(\mathfrak{a}), \qquad (\mathfrak{a}, F/K) = 1. \tag{18}$$

Note that both $\Lambda(\mathfrak{a})$ and $\varphi(\mathfrak{a})$ are now independent of any choice. To prove (18) observe that $\Lambda$ is multiplicative on the group of ideals for which $(\mathfrak{a}, F/K) = 1$. If $\varphi'$ is any extension of it to a grossencharacter, $\varphi' \circ N_{F/K} = \Lambda \circ N_{F/K} = \psi$

so $\varphi'$ is one of the $\varphi$ satisfying (13). To sum up, $\varphi$ and $\Lambda$ are uniquely determined by $E$ and $F$ on $\{\mathfrak{a} \mid (\mathfrak{a}, F/K) = 1\}$. From there on they can either be extended as a cocycle $\Lambda$, unique modulo coboundaries, or as a character $\varphi$, unique modulo $\widehat{Gal(F/K)}$.

For an interpretation of $\varphi$ by means of the abelian variety $Res_{F/K}E$, see [Go-Sch].

Finally, let $L$ be the period lattice of $\omega$, and

$$(19) \qquad \xi : \mathbf{C}/L \to E(\mathbf{C}) \qquad \xi(z, L) = (\wp(z, L), \wp'(z, L))$$

the corresponding analytic uniformization. Then

$$(20) \qquad L_{\mathfrak{a}} = \Lambda(\mathfrak{a})\mathfrak{a}^{-1}L$$

is the period lattice of $\omega^{\sigma_{\mathfrak{a}}}$ on $E^{\sigma_{\mathfrak{a}}}$, and the following diagram commutes ($\mathfrak{a}$ integral, $(\mathfrak{a}, \mathfrak{f}) = 1$) :

$$(21) \qquad \begin{array}{ccccc} \mathfrak{a}^{-1}L & \hookrightarrow & \mathbf{C}/L & \xrightarrow{\Lambda(\mathfrak{a})} & \mathbf{C}/L_{\mathfrak{a}} \\ \downarrow & & \downarrow \xi(\cdot, L) & & \downarrow \xi(\cdot, L_{\mathfrak{a}}) \\ Ker\lambda(\mathfrak{a}) & \hookrightarrow & E(\mathbf{C}) & \xrightarrow{\lambda(\mathfrak{a})} & E^{\sigma_{\mathfrak{a}}}. \end{array}$$

It follows that $g_k(L)^{\sigma_{\mathfrak{a}}} = g_k(L_{\mathfrak{a}})$ $(k = 2, 3)$ and $\Delta(L)^{\sigma_{\mathfrak{a}}} = \Delta(L_{\mathfrak{a}})$.

EXERCISE: Show that $\mathfrak{a}\mathcal{O}_F = \Lambda(\mathfrak{a})\mathcal{O}_F$.

**1.6 Proposition.** *Let $\mathfrak{g}$ be an integral ideal divisible by $\mathfrak{f}$. Then $K(\mathfrak{g}) = F(E[\mathfrak{g}])$.*

PROOF: Let $\mathfrak{A}$ be an integral ideal of $F$ prime to $\mathfrak{g}$ and suppose $\sigma_{\mathfrak{A}}(u) = u$ for every $u \in E[\mathfrak{g}]$. Since $\mathfrak{a} = N_{F/K}(\mathfrak{A}) = (a)$ with $a = \varphi(\mathfrak{a}) = \psi(\mathfrak{A})$, (11) implies that $a \equiv 1 \; mod \; \mathfrak{g}$. If $\mathcal{B}$ is an ideal of $F(E[\mathfrak{g}])$ and $\mathfrak{A}$ its norm down to $F$, then the argument above applies to $\mathfrak{A}$, so $N_{F(E[\mathfrak{g}])/K}\mathcal{B} = (a)$, $a \equiv 1 \; mod \; \mathfrak{g}$. By class field theory, $K(\mathfrak{g}) \subset F(E[\mathfrak{g}])$. Conversely, $F \subset K(\mathfrak{g})$, and if $\mathfrak{a} = (a)$, $a \equiv 1 \; mod \; \mathfrak{g}$, then $\lambda(\mathfrak{a}) = \iota(\varphi(\mathfrak{a})) = \iota(a)$, so by (15) $\sigma_{\mathfrak{a}}$ fixes $E[\mathfrak{g}]$ pointwise, showing $F(E[\mathfrak{g}]) \subset K(\mathfrak{g})$.

**1.7 Corollary.** *Let $\mathfrak{g}$ and $\mathfrak{m}$ be any two relatively prime integral ideals, $(\mathfrak{f}, \mathfrak{m}) = 1$. Then $F(E[\mathfrak{m}])$ and $F(E[\mathfrak{g}])$ are linearly disjoint over $F$, and $Gal(F(E[\mathfrak{m}])/F) \cong (\mathcal{O}_K/\mathfrak{m})^x$.*

PROOF: Without loss of generality, we may assume $\mathfrak{f}|\mathfrak{g}$, because the hypotheses are not changed if we replace $\mathfrak{g}$ by $\mathfrak{f}\mathfrak{g}$. Clearly $Gal(F(E[\mathfrak{m}])/F) \subset (\mathcal{O}_K/\mathfrak{m})^x$. Let $\Phi(\mathfrak{m}) = \sharp(\mathcal{O}_K/\mathfrak{m})^x$ (Euler's $\Phi$ function). By class field theory, $\Phi(\mathfrak{m}) = [K(\mathfrak{m}\mathfrak{g}) : K(\mathfrak{g})]$, since $(\mathfrak{m}, \mathfrak{g}) = 1$, and $w_{\mathfrak{g}} = 1$. The corollary now follows from this and from proposition 1.6. The field diagram is described below.

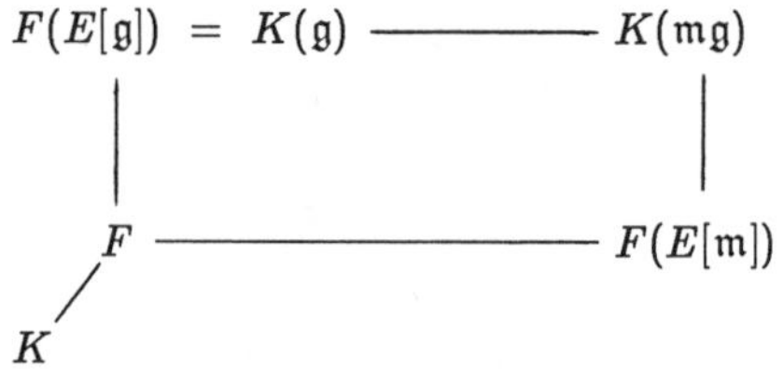


**1.8** GOOD AND BAD REDUCTION: Let $\mathfrak{P}$ be a prime of $F$. $E$ has *good reduction* at $\mathfrak{P}$ if there exists an elliptic curve $\mathcal{E}$ over the localization $R$ of $\mathcal{O}_F$ at $\mathfrak{P}$, whose generic fiber is isomorphic to $E$ over $F$. The special fiber $\mathcal{E} \times Spec\ \mathcal{O}_F/\mathfrak{P}$ is then denoted by $\tilde{E}$, the *reduction* of $E$ *mod* $\mathfrak{P}$. A basic theorem asserts that $\mathcal{E}$, hence $\tilde{E}$, depend only on $E$. It is also known that if $E$ has good reduction at $\mathfrak{P}$, there exists a Weierstrass model over $R$ with $\Delta \in R^x$ (if the residual characteristic of $\mathfrak{P}$ is 2 or 3 we must allow for a generalized Weierstrass model; see 1.11 below). The corresponding differential of the first kind $\omega_E$ is then a basis for $H^0(\mathcal{E}, \Omega^1_{\mathcal{E}/R})$. Every $F$-rational point of $E$ extends uniquely to an $R$-rational point of $\mathcal{E}$ (since $\mathcal{E}/R$ is smooth and proper) and when we read it modulo $\mathfrak{P}$ we get the *reduction map* $E(F) \rightarrow \tilde{E}(\mathcal{O}_F/\mathfrak{P})$. The kernel of reduction *mod* $\mathfrak{P}$ will be written $E_{1,\mathfrak{P}}$. In a Weierstrass model (6) with $\Delta \in R^x$, $E_{1,\mathfrak{P}}$ consists of all the points with non-integral affine coordinates.

The following fundamental theorem is proved in [Se-Ta] as a consequence of the theory of Néron models.

**Theorem.** *(i) The primes of bad reduction for $E/F$ are precisely the primes dividing its conductor.*

*(ii) (Criterion of Ogg-Néron-Shafarevitch). Let $\mathfrak{m}$ be an integral ideal of $K$, relatively prime to $\mathfrak{P}$. Then $\mathfrak{P}$ is a prime of good reduction if and only if $F(E[\mathfrak{m}^\infty])/F$ is unramified at $\mathfrak{P}$.*

**1.9** We shall now determine the decomposition patterns of primes of $F$ in the extensions obtained by adjoining division points of $E$.

**Proposition.** *Let $\mathfrak{p}$ be a prime of $K$, $(\mathfrak{p}, \mathfrak{f}) = 1$, and let $F_n = F(E[\mathfrak{p}^n])$, $0 \leq n \leq \infty$.*

*(i) All the primes above $\mathfrak{p}$ are totally ramified in $F_\infty/F$.*

*(ii) Primes not above $\mathfrak{p}$ are finitely ramified (i.e. their inertia group in $Gal(F_\infty/F)$ is finite), unramified if they are primes of good reduction.*

*(iii) If $\mathfrak{p}$ is split in $K/\mathbf{Q}$, every prime not above $\mathfrak{p}$ is finitely decomposed in $F_\infty/F$. If $\mathfrak{p}$ is inert (resp. ramified) and $\mathfrak{Q}$ is a prime of $F$ not above $\mathfrak{p}$, the order of the decomposition group of $\mathfrak{Q}$ in $Gal(F_n/F)$ is asymptotic to $cp^n$ (resp. $cp^{n/2}$) for large $n$, $c$ being a constant.*

PROOF: We shall deduce (i) in 1.10 from formal group considerations. To prove (ii) and (iii) we may replace $F$ by $K(\mathfrak{g})$, $\mathfrak{f}|\mathfrak{g}$, so that over $K(\mathfrak{g})$ $\psi = \varphi \circ N_{K(\mathfrak{g})/K}$ is unramified, and $E$ has good reduction everywhere. Then (ii) follows from 1.8. (Note that it is not enough to take $\mathfrak{g} = \mathfrak{f}$ in general).

To prove (iii) let $\mathfrak{Q}$ be a prime of $F$ not above $\mathfrak{p}$. Since we may assume now that $\mathfrak{Q}$ is unramified, its decomposition group in $F_n/F$ is cyclic, generated by Frobenius at $\mathfrak{Q}$. It follows from (11) that its order is the least positive integer $f$ such that

$$\psi(\mathfrak{Q})^f \equiv 1 \mod \mathfrak{p}^n \mathcal{O}_\mathfrak{p}. \tag{22}$$

The assertions in (iii) all follow from this.

**1.10** THE FORMAL GROUP: Let $(\mathfrak{p}, \mathfrak{f}) = 1$, and denote by $\mathfrak{P}$ a prime of $F$ above $\mathfrak{p}$. Fix a Weierstrass model of $E$ over $R$ (the localization of $\mathcal{O}_F$ at $\mathfrak{P}$) with $\Delta \in R^\times$.

Let $\hat{E}$ be the one-parameter formal group law of $E$ with respect to the parameter

(23) $$t = -\frac{2x}{y}$$

(see 1.11 for modifications in residual characteristic 2 or 3) ([Ta] §3). $\hat{E}$ is defined over $R$ but we consider it over the completion of $R$, $\mathcal{O}_{\mathfrak{P}}$.

**Lemma.** *$\hat{E}$ is a relative Lubin-Tate group with respect to the unramified extension $F_{\mathfrak{P}}/K_{\mathfrak{p}}$. It is of height 1 if $\mathfrak{p}$ is split in $K/\mathbf{Q}$, and of height 2 if $\mathfrak{p}$ is inert or ramified.*

PROOF: Let $\phi = \sigma_{\mathfrak{p}}$ be the Frobenius automorphism. The isogeny $\lambda(\mathfrak{p}) : E \to E^{\phi}$ induces a homomorphism of formal groups $\widehat{\lambda(\mathfrak{p})} : \hat{E} \to \hat{E}^{\phi}$ which is of the form

(24) $$\begin{cases} \widehat{\lambda(\mathfrak{p})}(t) & = \Lambda(\mathfrak{p})t + \ldots \in \mathcal{O}_{\mathfrak{P}}[[t]] \\ & \equiv t^q \; mod \; \mathfrak{P}\mathcal{O}_{\mathfrak{P}} \end{cases}$$

with $q = \mathbf{N}\mathfrak{P}$. See I.1.2. The rest is obvious in view of the results of chapter I.

**Corollary.** *$\mathfrak{P}$ is totally ramified in $F_{\infty} = F(E[\mathfrak{p}^{\infty}])$ (see 1.9(i)).*

PROOF: Comparing the representation of $Gal(F_{\infty}/F)$ in $\mathcal{O}_{\mathfrak{p}}^{x} = Aut(E[\mathfrak{p}^{\infty}])$ (see corollary 1.7) and the corresponding representation of the local Galois group in $\mathcal{O}_{\mathfrak{p}}^{x} = Aut(\hat{E}[\mathfrak{p}^{\infty}])$ ($\hat{E}[\mathfrak{p}^{\infty}]$ was denoted $W_f$ in chapter I), we see that the two images coincide. Therefore, $\mathfrak{P}$ does not decompose in $F_{\infty}$. But the local tower ($k_{\xi}/k'$ in the notation of chapter I) is totally ramified, so the corollary follows.

Another consequence that will be needed later is the following. Consider $\prod' t(Q)$ where the product is over non-zero $Q$ in $E[\mathfrak{p}]$. Since $F(E[\mathfrak{p}])/F$ is totally ramified at $\mathfrak{P}$, and the $\mathbf{N}\mathfrak{p} - 1$ points $Q$ are conjugate to each other, *the order of this product at $\mathfrak{P}$ is* 1. *This applies to each* $\mathfrak{P}|\mathfrak{p}$.

**1.11** CHARACTERISTIC 2 AND 3: Some formulas concerning Weierstrass models have to be modified when the residual characteristic is 2 or 3. See [Ta] for details. A *generalized Weierstrass model* is given by

(6') $$y^2 + a_1xy + a_3y = x^3 + a_2x^2 + a_4x + a_6.$$

The $\Delta$ function is given by [Ta](2). With these, the assertions made in 1.8 about good reduction and Weierstrass models remain valid. The differential $\omega_E$ becomes

$$\omega_E = \frac{dx}{2y + a_1x + a_3} = \frac{dy}{3x^2 + 2a_2x + a_4 - a_1y}, \tag{10'}$$

and the lattice $L$ is still its period lattice. The uniformization $\xi : \mathbf{C}/L \to E(\mathbf{C})$ is accomplished through

$$\begin{cases} x = \wp(z, L) - (a_1^2 + 4a_2)/12 \\ y = (\wp'(z, L) - a_1x - a_3)/2. \end{cases} \tag{19'}$$

The parameter $t$ for the formal group is now given by

$$t = -\frac{x}{y} \tag{23'}$$

and with it $\hat{E}$ and $\omega_E$ are defined over $\mathbf{Z}[a_1, \ldots, a_6]$. The kernel of reduction *mod* $\mathfrak{P}$, $E_{1,\mathfrak{P}}$, is still the subgroup of points where $x$ and $y$ are non-integral. Restricted to $E_{1,\mathfrak{P}}$ $t$ is finite and $|t|_{\mathfrak{P}} < 1$.

## 2. Elliptic Units

Elliptic units are units in abelian extensions of quadratic imaginary fields, obtained as special values of elliptic modular functions. They play a role analogous to that of circular units in abelian number fields. See the works of C.L. Siegel [Sie], Ramachandra [Ra], Robert [R], Gillard and Robert [Gi-R], and Kubert and Lang [K-La]. Except for nineteenth-century results on elliptic functions (out of the many references here we mention Whittaker and Watson [W-W], Chandrasekharan [Chand], and Weil's beautiful little book [We2]), our exposition is self contained, and some of the proofs, I believe, are new.

**2.1** THETA FUNCTIONS: Let $L = \mathbf{Z}\omega_1 + \mathbf{Z}\omega_2$ be a lattice in $\mathbf{C}$, whose basis is ordered so that $\tau = \omega_1/\omega_2$ belongs to the upper half plane. Recall that Weierstrass'

$\sigma$ function and Ramanujan's $\Delta$ function have the absolutely convergent product expansions

$$(1) \qquad \sigma(z,L) = z \cdot \prod_{\omega\in L}{}' \left(1 - \frac{z}{\omega}\right) exp\left(\frac{z}{\omega} + \frac{1}{2}\left(\frac{z}{\omega}\right)^2\right)$$

$$(2) \qquad \Delta(L) = (2\pi i/\omega_2)^{12}\, q_\tau \prod_{\nu=1}^{\infty} (1 - q_\tau^\nu)^{24}, \quad q_\tau = e^{2\pi i\tau}.$$

$\sigma(z,L)$ satisfies the transformation law

$$(3) \qquad \sigma(z+\omega,L) = \pm\sigma(z,L) exp(\eta(\omega,L)(z+\frac{\omega}{2})), \quad \omega \in L,$$

where $\eta$ is an **R**-linear form on **C**, explicitly given by the following set of formulas.

$$(4) \qquad \begin{cases} \eta(z,L) = \dfrac{\omega_1\eta_2 - \omega_2\eta_1}{2\pi i A(L)}\,\overline{z} + \dfrac{\overline{\omega}_2\eta_1 - \overline{\omega}_1\eta_2}{2\pi i A(L)}\, z \\ A(L) = (2\pi i)^{-1}(\omega_1\overline{\omega}_2 - \overline{\omega}_1\omega_2) = \pi^{-1}\,\mathrm{Area}(\mathbf{C}/L) \\ \eta_1 = \omega_1 \sum_n \sum_m' (m\omega_1 + n\omega_2)^{-2}, \quad \eta_2 = \omega_2 \sum_m \sum_n' (m\omega_1 + n\omega_2)^{-2}. \end{cases}$$

The order of summation matters. Half way in between $\eta_1/\omega_1$ and $\eta_2/\omega_2$ we find

$$(5) \qquad s_2(L) = \lim_{0<s\to 0} \sum_{\omega\in L}{}' \omega^{-2}|\omega|^{-2s}.$$

With these definitions,

$$(6) \qquad \begin{cases} \omega_1\eta_2 - \omega_1\omega_2 \cdot s_2(L) = \dfrac{\overline{\omega}_2\omega_1}{A(L)} \\ \omega_2\eta_1 - \omega_1\omega_2 \cdot s_2(L) = \dfrac{\overline{\omega}_1\omega_2}{A(L)} \end{cases}$$

from which one recovers Legendre's relation $\omega_1\eta_2 - \omega_2\eta_1 = 2\pi i$, and also (using (4)) $\eta(z,L) = A(L)^{-1}\overline{z} + s_2(L)z$.

The *fundamental theta function*

$$(7) \qquad \theta(z,L) = \Delta(L) \cdot e^{-6\eta(z,L)z} \cdot \sigma(z,L)^{12}$$

is non-holomorphic, but its arithmetical usefulness makes up for this blemish. If $c \neq 0$, $\theta(cz, cL) = \theta(z, L)$ (the reason for $\Delta(L)$ in (7)), and the exponential is chosen so that $|\theta(z, L)|$ is $L$-periodic. In addition, $\theta$ posseses an important product expansion. Normalize $L$ so that $\omega_1 = \tau$, $\omega_2 = 1$, and let $q_z = e^{2\pi i z}$. Then ([We2], IV §3(15))

$$(8)\quad \theta(z, L) = e^{6A(L)^{-1}z(z-\bar{z})} \cdot q_\tau (q_z^{1/2} - q_z^{-1/2})^{12} \cdot \prod_{\nu=1}^{\infty} \{(1 - q_\tau^\nu q_z)(1 - q_\tau^\nu q_z^{-1})\}^{12}.$$

Many identities, and in particular (11) below, are consequences of (8).

**2.2** SIEGEL UNITS IN THE HILBERT CLASS FIELD: We shall have almost no use for these units, but since their construction is so simple, we describe it briefly.

Let $K$ be a quadratic imaginary field. For any ideal $\mathfrak{a}$ of $K$, consider

$$(9)\qquad u(\mathfrak{a}) = \frac{\Delta(\mathcal{O}_K)}{\Delta(\mathfrak{a}^{-1}\mathcal{O}_K)}.$$

**Proposition.** *(i)* $u(\mathfrak{a}) \in K(1)$,

*(ii)* $u(\mathfrak{a}\mathfrak{b}) = u(\mathfrak{a})^{\sigma_\mathfrak{b}} \cdot u(\mathfrak{b})$,

*(iii)* $(u(\mathfrak{a})) = \mathfrak{a}^{-12}\mathcal{O}_{K(1)}$.

PROOF: See [Sie], chapter II, §2.

If $h$ is the class number of $K$, then $\mathfrak{a}^h = (\alpha)$, $\alpha \in K^x$, and although $\alpha$ is not unique, $\alpha^{12}$ is, so $\delta(\mathfrak{a}) = u(\mathfrak{a})^h \alpha^{12}$ is a well defined unit in $K(1)$. The group of Siegel units in $K(1)$ is generated by the $\delta(\mathfrak{a})$. It is of finite index in the group of all units, and stable under Galois.

**2.3** From now on assume that $L$ admits complex multiplications by $\mathcal{O}_K$. Let $\mathfrak{a}$ be an integral ideal of $K$. The function

$$(10)\qquad \begin{cases} \Theta(z; L, \mathfrak{a}) & = \theta(z, L)^{\mathrm{N}\mathfrak{a}} / \theta(z, \mathfrak{a}^{-1}L) \\ & = \dfrac{\Delta(L)}{\Delta(\mathfrak{a}^{-1}L)} \cdot \prod'_{u \in \mathfrak{a}^{-1}L/L} \dfrac{\Delta(L)}{(\wp(z, L) - \wp(u, L))^6} \end{cases}$$

is an elliptic function with respect to $L$ (to obtain the second expression in (10) compare divisors and leading terms in the Taylor expansions).

**Proposition** (THE DISTRIBUTION RELATION). *Let $\mathfrak{a}$ and $\mathfrak{b}$ be integral ideals of $K$, relatively prime to each other. Then*

$$\prod_{v \in \mathfrak{b}^{-1}L/L} \Theta(z+v; L, \mathfrak{a}) = \Theta(z; \mathfrak{b}^{-1}L, \mathfrak{a}). \tag{11}$$

PROOF: Both sides of (11) are $L$-elliptic functions with the same divisor, so their ratio is a constant. Comparing the Taylor expansions at $z = 0$ shows that this constant is

(12)
$$\varepsilon(\mathfrak{a},\mathfrak{b}) = \left(\frac{\Delta(L)}{\Delta(\mathfrak{a}^{-1}L)}\right)^{\mathrm{N}\mathfrak{b}-1} \cdot \left(\frac{\Delta(L)}{\Delta(\mathfrak{b}^{-1}L)}\right)^{\mathrm{N}\mathfrak{a}-\sigma_\mathfrak{a}} \cdot \prod_u{}' \prod_v{}' \frac{\Delta(L)}{(\wp(u,L)-\wp(v,L))^6}.$$

Here $u$ runs over $\mathfrak{a}^{-1}L/L - \{0\}$, $v$ over $\mathfrak{b}^{-1}L/L - \{0\}$, and the denominator never vanishes since $(\mathfrak{a},\mathfrak{b}) = 1$. Despite the apparent asymmetry, $\varepsilon(\mathfrak{a},\mathfrak{b}) = \varepsilon(\mathfrak{b},\mathfrak{a})$, because $(\Delta(L)/\Delta(\mathfrak{b}^{-1}L))^{\sigma_\mathfrak{a}-1} = (\Delta(L)/\Delta(\mathfrak{a}^{-1}L))^{\sigma_\mathfrak{b}-1}$ (2.2(ii)). We shall prove that $\varepsilon(\mathfrak{a},\mathfrak{b}) = 1$.

Let $H = K(1)$ be the Hilbert class field of $K$, and $w_H$ the number of roots of unity in it. A short analysis of ramification patterns reveals that $w_H|12$. In fact $4|w_H$ if and only if $d_K \equiv 4 \mod 8$, and $6|w_H$ if and only if $d_K \equiv 0 \mod 3$. Now $\varepsilon(\mathfrak{a},\mathfrak{b})$ evidently *lies in* $H$, because we may choose $L$ so that the particular Weierstrass model is defined over $H$. Next we claim that $\varepsilon(\mathfrak{a},\mathfrak{b})$ *is a root of unity.* The proof of this fact is a tedious but straightforward computation, based on (8). It can be found in [K-La] chapter 2, theorem 4.1(i), from where it actually follows that $\varepsilon(\mathfrak{a},\mathfrak{b})^{\mathrm{N}\mathfrak{b}} = 1$. Thanks to the symmetry between $\mathfrak{a}$ and $\mathfrak{b}$, $\varepsilon(\mathfrak{a},\mathfrak{b})^{\mathrm{N}\mathfrak{a}} = 1$.

So far we know that $\varepsilon(\mathfrak{a},\mathfrak{b})^m = 1$ where $m = g.c.d.(w_H, \mathrm{N}\mathfrak{a}, \mathrm{N}\mathfrak{b})$. To conclude we need the following lemma of Robert ([K-La] chapter 11, 5.7).

**Lemma**. *Suppose $g_k(L) \in H$, $k = 2,3$. Then for any $\mathfrak{a}$ relatively prime to $w_K$,*

$$\Delta(L)^{\mathrm{N}\mathfrak{a}}/\Delta(\mathfrak{a}^{-1}L) \in (H^x)^{12}.$$

Consider the expression (12) for $\varepsilon(\mathfrak{a},\mathfrak{b})$, where $L$ is chosen so that $g_k(L) \in H$. Since $\wp(u) = \wp(-u)$, $\prod'\prod'(\wp(u) - \wp(v))^6$ is a twelfth power in $H^x$, unless both

$\mathfrak{a}$ and $\mathfrak{b}$ are even (not relatively prime to 2). But this can only happen if 2 splits in $K$, and then $w_H|6$. Therefore, in any case

$$(13)\qquad \varepsilon(\mathfrak{a},\mathfrak{b}) \equiv \left(\frac{\Delta(L)^{\mathrm{N}\mathfrak{a}}}{\Delta(\mathfrak{a}^{-1}L)}\right)^{\mathrm{N}\mathfrak{b}-1} \left(\frac{\Delta(L)}{\Delta(\mathfrak{b}^{-1}L)}\right)^{\mathrm{N}\mathfrak{a}-\sigma_\mathfrak{a}} \quad mod\ (H^x)^{w_H}.$$

Choose $\beta \in K^x$ so that $\mathfrak{b}' = (\beta)\mathfrak{b}$ is integral and $(\mathfrak{b}', 6\mathfrak{a}) = 1$. Then (13) implies

$$(14)\qquad \varepsilon(\mathfrak{a},\mathfrak{b}) \equiv \left(\frac{\Delta(L)^{\mathrm{N}\mathfrak{a}}}{\Delta(\mathfrak{a}^{-1}L)}\right)^{\mathrm{N}\mathfrak{b}-\mathrm{N}\mathfrak{b}'} \quad mod\ (H^x)^{w_H},$$

because $\varepsilon(\mathfrak{a},\mathfrak{b}') = 1$. From Robert's lemma we conclude that $\varepsilon(\mathfrak{a},\mathfrak{b}) = 1$ if $(\mathfrak{a}, w_K) = 1$.

Suppose $K$ is not $\mathbf{Q}(i)$ or $\mathbf{Q}(\sqrt{-3})$. If $\mathfrak{a}$ or $\mathfrak{b}$ is odd, $\varepsilon(\mathfrak{a},\mathfrak{b}) = 1$. If both $\mathfrak{a}$ and $\mathfrak{b}$ are even, 2 splits in $K$, so $w_H = 2$ or 6. Pick $\mathfrak{b}' = (\beta)\mathfrak{b}$ such that $(\mathfrak{b}', 2\mathfrak{a}) = 1$ and $\mathrm{N}\mathfrak{b} \equiv \mathrm{N}\mathfrak{b}'$ *mod* 3. Then if $w_H = 6$, (14) shows that $\varepsilon(\mathfrak{a},\mathfrak{b})$ is a cube. If we also show that $\varepsilon(\mathfrak{a},\mathfrak{b})$ is a square, it would belong to $(H^x)^{w_H}$. But then $\varepsilon(\mathfrak{a},\mathfrak{b}) = 1$. Formula (14) implies $\varepsilon(\mathfrak{a},\mathfrak{b}) \equiv \Delta(L)$ *mod* $(H^x)^2$. Now, with the model $y^2 = 4x^3 - hx - h$, $h = 27j/(j-1728)$, for the elliptic curve with $j$-invariant $j$, $\Delta \equiv j - 1728$ *mod* $(H^x)^2$. A theorem of Weber ([Web] §§134-135) asserts that $j - 1728$ is a square in $H$, and concludes the proof whenever $K$ is not $\mathbf{Q}(i)$ or $\mathbf{Q}(\sqrt{-3})$.

Finally, in the two exceptional cases $K = H$, and $\mathfrak{a}$ and $\mathfrak{b}$ are principal. If $K = \mathbf{Q}(i)$ either $\mathfrak{a}$ or $\mathfrak{b}$ must be odd, so $\varepsilon(\mathfrak{a},\mathfrak{b}) = 1$ is a consequence of (14) and Robert's lemma, as above. If $K = \mathbf{Q}(\sqrt{-3})$, either $\mathfrak{a}$ or $\mathfrak{b}$ is odd, and one of them is also prime to 3. From (13) we get

$$\varepsilon(\mathfrak{a},\mathfrak{b}) \equiv \Delta(L)^{(\mathrm{N}\mathfrak{a}-1)(\mathrm{N}\mathfrak{b}-1)} \ mod\ (H^x)^6$$

and $6|(\mathrm{N}\mathfrak{a}-1)(\mathrm{N}\mathfrak{b}-1)$, so again $\varepsilon(\mathfrak{a},\mathfrak{b})$ is a sixth power, hence 1. The proof of the proposition is now complete.

**2.4 Proposition.** *Let $\mathfrak{m}$ be a non-trivial integral ideal of $K$, and $v$ a primitive $\mathfrak{m}$-division point of $L$ (i.e. $v \in \mathfrak{m}^{-1}L$, but $v \notin \mathfrak{n}^{-1}L$ for any proper divisor $\mathfrak{n}$ of $\mathfrak{m}$). Then, if $(\mathfrak{a},\mathfrak{m}) = 1$,* *(i)* $\Theta(v; L, \mathfrak{a}) \in K(\mathfrak{m})$.

*(ii)* $\Theta(v; L, \mathfrak{a})^{\sigma_{\mathfrak{c}}} = \Theta(v; \mathfrak{c}^{-1}L, \mathfrak{a}) = \Theta(v; L, \mathfrak{ac})\Theta(v; L, \mathfrak{c})^{-\mathbf{N}\mathfrak{a}}$ *($\mathfrak{c}$ integral, relatively prime to $\mathfrak{m}$).*

*(iii)* $\Theta(v; L, \mathfrak{a})$ *is a unit if $\mathfrak{m}$ is not a prime power. If $\mathfrak{m} = \mathfrak{p}^n$, it is a unit outside $\mathfrak{p}$.*

PROOF: (i) We may assume that $L$ corresponds to a Weierstrass model $E$ defined over the Hilbert class field $K(1)$. Part (i) follows from (10) and standard results from the theory of complex multiplication ([Sh], theorem 5.5). It is also a consequence of (ii).

(ii) Once more we are free to choose a model $E$ for $\mathbf{C}/L$, changing $L$ by a homothety, because $\Theta(v; L, \mathfrak{a})$ depends only on the *complex* isomorphism class of the elliptic curve. Choose an integral ideal $\mathfrak{f}$ with $w_{\mathfrak{f}} = 1$, relatively prime to $\mathfrak{c}$, and assume that the model $E$ corresponding to $L$ is defined over $K(\mathfrak{f}) = F$, and satisfies $\psi_{E/F} = \varphi \circ N_{F/K}$ for a grossencharacter $\varphi$ whose conductor divides $\mathfrak{f}$. Then (10) and 1.5(21) yield $\Theta(v; L, \mathfrak{a})^{\sigma_{\mathfrak{c}}} = \Theta(\Lambda(\mathfrak{c})v; \Lambda(\mathfrak{c})\mathfrak{c}^{-1}L, \mathfrak{a}) = \Theta(v; \mathfrak{c}^{-1}L, \mathfrak{a})$. The rest of (ii) follows from this, and from $\Theta(v; L, \mathfrak{a}) = \theta(v, L)^{\mathbf{N}\mathfrak{a}}/\theta(v, \mathfrak{a}^{-1}L)$.

(iii) We shall give two different proofs of this important result.

FIRST PROOF: We make the same assumption as in (ii) on the model $E$, and in addition we assume $(\mathfrak{f}, \mathfrak{m}) = 1$. Suppose $\mathfrak{p}|\mathfrak{m}$ and let $M_n = F(E[\mathfrak{m}\mathfrak{p}^n])$, $n \geq 0$. Corollary 1.7 implies that the conjugates of a primitive $\mathfrak{m}\mathfrak{p}^n$-division point $v$ over $M_{n-1}$ $(n \geq 1)$ are $v + u$ for $u \in E[\mathfrak{p}]$. Let $e_n = \Theta(v; \mathfrak{p}^n L, \mathfrak{a})$. Although $e_n = \Theta(\Lambda(\mathfrak{p}^{-n})v; \Lambda(\mathfrak{p}^{-n})\mathfrak{p}^n L, \mathfrak{a})$ refers to $\sigma_{\mathfrak{p}}^{-n}(E)$, and not to $E$, it lies in $M_n$, because all the conjugates of $E$ are $F$-isogenous, so the fields $F(E^{\sigma}[\mathfrak{c}])$ coincide for all $\sigma \in Gal(F/K)$. Since $v$ is a primitive $\mathfrak{m}\mathfrak{p}^n$-division point of $\mathfrak{p}^n L$, the distribution relation (11) implies (with $\mathfrak{b} = \mathfrak{p}$) $N_{n,n-1}e_n = e_{n-1}$. Here $N_{m,n}$ is the norm from $M_m$ to $M_n$.

Suppose first that $\mathfrak{p}$ is a split prime. According to 1.9, only primes above $\mathfrak{p}$ may ramify in $M_\infty/M_0$, and all the others are *finitely decomposed* in that tower. Let $\mathfrak{q}$ be a prime of $K$ different from $\mathfrak{p}$, and choose $n$ large enough so that all the prime divisors of $\mathfrak{q}$ are inert in $M_\infty/M_n$. Since $e_n \in N_{m,n}M_m^x$, $m \geq n$, $e_n$ is a unit above $\mathfrak{q}$, hence so is $e_0 = N_{n,0}e_n$.

If $\mathfrak{p}$ is inert or ramified this argument needs a slight modification. First, replacing $\mathfrak{m}$ by $\mathfrak{m}\mathfrak{p}^n$, $e_0$ by $e_n$, we may assume that $Gal(M_\infty/M_0) \cong \mathbf{Z}_p^2$ (a redundant step if $(\mathfrak{p}, 6) = 1$!). Proposition 1.9 implies that the decomposition group $D$ of any prime of $M_0$ above $\mathfrak{q}$ (they all coincide) is $\mathbf{Z}_p$. Once again replacing $M_0$ by some $M_n$, we shall assume that $Gal(M_\infty/M_0)/D \cong \mathbf{Z}_p$ too. It follows that $M_\infty$ contains a $\mathbf{Z}_p$ extension $N_\infty$ of $M_0$ in which every prime above $\mathfrak{q}$ is inert. The rest is identical to the split case, because $e_0$ sits in a norm-coherent sequence in the tower $N_\infty$.

We conclude that $\Theta(v; L, \mathfrak{a})$ is a unit outside $\mathfrak{p}$, therefore if $\mathfrak{m}$ is divisible by two distinct primes, a unit.

REMARK: If $\mathfrak{m} = \mathfrak{p}^n$, $\Theta(v; L, \mathfrak{a})$ is *not* a unit. The reader should keep in mind the cyclotomic analogy: $exp(2\pi i/m) - 1$ is a unit if and only if $m$ is divisible by two distinct primes.

SECOND PROOF: This proof does not use the distribution relation (11). It does not even use complex multiplication, and works whenever $j(L)$ is an algebraic integer, or even more generally, if one is willing to use some results about Tate curves. This interesting observation, that the elliptic units "are units" even in the absence of complex multiplication, does not seem to have found yet far reaching applications.

We may assume that the lattice $L$ corresponds to an elliptic curve $E$ defined over a number field $F$, with good reduction everywhere, and that $E[\mathfrak{m}\mathfrak{a}]$ are rational over $F$. Fix a prime $\mathfrak{Q}$ of $F$, and a Weierstrass model over $R$, the localization of $\mathcal{O}_F$ at $\mathfrak{Q}$, with $\Delta \in R^x$. We shall use $\sim$ to mean "both sides have the same valuation at $\mathfrak{Q}$". Let $P = \xi(v, L)$ be the point corresponding to $v$, and $E_1$ the kernel of reduction modulo $\mathfrak{Q}$. Thus $Q \in E_1$ if and only if $|x(Q)|_{\mathfrak{Q}} > 1$. Suppose first that $\mathfrak{a}$ is *prime* and recall (1.10) that

$$\prod_{Q \in E[\mathfrak{a}]}{}' t(Q) \sim \mathfrak{a} \qquad \text{if } \mathfrak{Q}|\mathfrak{a}. \tag{15}$$

Since $\Delta(L) \sim 1$, (10) implies

$$\Theta(v; L, \mathfrak{a}) \sim \mathfrak{a}^{-12} \prod_{Q \in E[\mathfrak{a}]}{}' (x(Q) - x(P))^{-6}. \tag{16}$$

Now suppose that either $\mathfrak{m}$ is divisible by two distinct primes, or else $\mathfrak{Q} \nmid \mathfrak{m}$. In both cases $P$, which is a primitive $\mathfrak{m}$ torsion point, does not belong to $E_1$. If $\mathfrak{Q} \nmid \mathfrak{a}$, then any non-zero $Q \in E[\mathfrak{a}]$ lies outside $E_1$, and so do $Q \pm P$. Therefore, $x(P)$ and $x(Q)$ are $\mathfrak{Q}$-integral and *non-congruent* modulo $\mathfrak{Q}$, whereby (16) is a unit at $\mathfrak{Q}$. If, on the other hand, $\mathfrak{Q}|\mathfrak{a}$, then for any $Q \in E[\mathfrak{a}] - \{0\}$, $x(Q) - x(P) \sim x(Q) \sim t(Q)^{-2}$, and (16) $\sim$ 1 because of (15). We conclude that as long as $\mathfrak{a}$ is prime and $(\mathfrak{a}, \mathfrak{m}) = 1$, (iii) holds. The general case follows at once, since for integral $\mathfrak{a}$, $\mathfrak{b}$, $\Theta(v; L, \mathfrak{a}\mathfrak{b}) = \Theta(v; L, \mathfrak{a})^{\mathbf{N}\mathfrak{b}} \cdot \Theta(v; \mathfrak{a}^{-1}L, \mathfrak{b})$.

EXERCISE: If $\mathfrak{m} = \mathfrak{p}^n$, and $\mathfrak{P}|\mathfrak{p}$, show that the $\mathfrak{P}$-adic valuation of $\Theta(v; L, \mathfrak{a})$ is $(\mathbf{N}\mathfrak{a} - 1) \times$ (a constant depending on $\mathfrak{m}$).

**2.5 Proposition.** *(i) Let $\mathfrak{f}$ be a non-trivial integral ideal of $K$, and $\mathfrak{g} = \mathfrak{f}\mathfrak{l}$, with a prime $\mathfrak{l}$. Let $e = w_\mathfrak{f}/w_\mathfrak{g}$. Then if $v$ is a primitive $\mathfrak{f}$-division point of $L$, and $(\mathfrak{a}, \mathfrak{g}) = 1$,*

$$N_{K(\mathfrak{g})/K(\mathfrak{f})}\Theta(v; \mathfrak{l}L, \mathfrak{a})^e = \begin{cases} \Theta(v; L, \mathfrak{a})^{1-\sigma_\mathfrak{l}^{-1}} & \text{if } \mathfrak{l} \nmid \mathfrak{f} \\ \Theta(v; L, \mathfrak{a}) & \text{if } \mathfrak{l} \mid \mathfrak{f}. \end{cases}$$

*(ii) Let $\mathfrak{l}$ be prime, $(\mathfrak{a}, \mathfrak{l}) = 1$ and $e = w_K/w_\mathfrak{l}$. Then if $v$ is a primitive $\mathfrak{l}$-division point of $L$, and $L$ represents the trivial class in $Pic(\mathcal{O}_K)$,*

$$N_{K(\mathfrak{l})/K(1)}\Theta(v; L, \mathfrak{a})^e = u(\mathfrak{l})^{\sigma_\mathfrak{a} - \mathbf{N}\mathfrak{a}}$$

*(see 2.2).*

PROOF: (i) We have seen that $[K(\mathfrak{g}) : K(\mathfrak{f})] = w_\mathfrak{g}\Phi(\mathfrak{g})/w_\mathfrak{f}\Phi(\mathfrak{f})$, where $\Phi(\mathfrak{f}) = |(\mathcal{O}_K/\mathfrak{f})^\times|$. Consider first the case $\mathfrak{l}|\mathfrak{f}$. The conjugates of $\Theta(v; \mathfrak{l}L, \mathfrak{a})$ over $K(\mathfrak{f})$ are $\Theta(v + u; \mathfrak{l}L, \mathfrak{a})$, $u \in L/\mathfrak{l}L$, and when $u$ ranges over the $\mathbf{N}\mathfrak{l}$ possible points, each of them is counted $e$ times. Our formula follows from the distribution relation (10).

If $\mathfrak{l} \nmid \mathfrak{f}$, then the conjugates of $\Theta(v; \mathfrak{l}L, \mathfrak{a})$ are $\Theta(v+u; \mathfrak{l}L, \mathfrak{a})$ for the $\mathbf{N}\mathfrak{l} - 1$ points $u \in L/\mathfrak{l}L$ for which $v+u$ is of level $\mathfrak{g}$. Again each conjugate appears $e$ times. If $u_0$ is the unique point such that $v+u_0$ is an $\mathfrak{f}$-division point of $\mathfrak{l}L$, then from 2.4(iii)

$$\Theta(v+u_0; \mathfrak{l}L, \mathfrak{a})^{\sigma_\mathfrak{l}} = \Theta(v+u_0; L, \mathfrak{a}) = \Theta(v; L, \mathfrak{a}).$$

As before, (10) gives the desired formula

(ii) This is proved in a similar way, and is left to the reader.

**2.6** ROBERT UNITS: We want to briefly indicate the relation between $\Theta(v; L, \mathfrak{a})$ and Robert's units.

Let $\mathfrak{f}$ be a *non-trivial* ideal of $K$, and $f$ the least positive integer in $\mathfrak{f} \cap \mathbf{Z}$. Let $Cl(\mathfrak{f})$ be the ray class group *mod* $\mathfrak{f}$, identified with $G(\mathfrak{f}) = Gal(K(\mathfrak{f})/K)$. Let $J(\mathfrak{f}) \subset \mathbf{Z}[G(\mathfrak{f})]$ be the ideal generated by $\sigma_\mathfrak{a} - \mathbf{N}\mathfrak{a}$, $(\mathfrak{a}, 6\mathfrak{f}) = 1$. $J(\mathfrak{f})$ is the annihilator of $\mu_{K(\mathfrak{f})}$. For any $\sigma \in G(\mathfrak{f})$, Robert's invariant is defined as

(17) $$\varphi_\mathfrak{f}(\sigma) = \theta(1, \mathfrak{f}\mathfrak{c}^{-1})^f, \qquad \sigma = (\mathfrak{c}, K(\mathfrak{f})/K), \ \mathfrak{c} \text{ integral.}$$

This is well defined, and the symbol $\varphi_\mathfrak{f}$ extends linearly to $\mathbf{Z}[G(\mathfrak{f})]$.

**Proposition** ([R], [GI-R]). *(i)* $\varphi_\mathfrak{f}(u) \in K(\mathfrak{f})$.

*(ii)* $\sigma\varphi_\mathfrak{f}(u) = \varphi_\mathfrak{f}(\sigma u)$.

*(iii) If $\mathfrak{f}$ is divisible by two distinct primes, or if $u$ belongs to the augmentation ideal, $\varphi_\mathfrak{f}(u)$ is a unit. If $\mathfrak{f} = \mathfrak{p}^n$, it is a $\mathfrak{p}$-unit.*

*(iv) Assume that $\mathfrak{f}$ is the conductor of $K(\mathfrak{f})$. If $u \in J(\mathfrak{f})$, $\varphi_\mathfrak{f}(u)$ is a $(12fw_\mathfrak{f})$th power in $K(\mathfrak{f})$ ([Gi-R] proposition A-2).*

Now the relation with 2.5 is given by

(18) $$\varphi_\mathfrak{f}(\mathbf{N}\mathfrak{a} - \sigma_\mathfrak{a}) = \Theta(1; \mathfrak{f}, \mathfrak{a})^f, \qquad \mathfrak{f} \neq (1).$$

The moral is that to get well defined units from $\theta(z, L)$ evaluated at a primitive $\mathfrak{f}$-division point, we may either raise to power $f$, or twist by $\sigma_\mathfrak{a} - \mathbf{N}\mathfrak{a}$, and the two operations are related by (18).

**2.7** The following result is similar to 2.6(iv), and will be used in 4.12 to deal with $p$-adic $L$ functions for $p = 2, 3$.

**Proposition.** *Let $v$ be a primitive $\mathfrak{f}$-division point of $L$ $(\mathfrak{f} \neq (1))$. Then for* $(\mathfrak{a}, 6\mathfrak{f}) = 1$, $\Theta(v; L, \mathfrak{a})$ *is a $12w_\mathfrak{f}$ power in $K(\mathfrak{f})^x$.*

We omit the proof. The proposition allows us finally to define the group of *primitive Robert units of conductor* $\mathfrak{f}$, denoted $C_\mathfrak{f}$.

DEFINITION: Let $\mathfrak{f}$ be a non-trivial integral ideal. Let $\Theta_\mathfrak{f}$ be the subgroup of $K(\mathfrak{f})^x$ generated by $\Theta(1; \mathfrak{f}, \mathfrak{a})$, $(\mathfrak{a}, 6\mathfrak{f}) = 1$. Define $C_\mathfrak{f}$ to be the group of all units in $K(\mathfrak{f})^x$ whose $12w_\mathfrak{f}$ power lies in $\mu_{K(\mathfrak{f})}\Theta_\mathfrak{f}$.

Notice that $C_\mathfrak{f}$ is Galois stable, that $\Theta_\mathfrak{f} \subset C_\mathfrak{f}$ only if $\mathfrak{f}$ is divisible by two distinct primes, and that from Robert's point of view, $C_\mathfrak{f}$ consists of all the units in $K(\mathfrak{f})$ whose $12fw_\mathfrak{f}$ power belongs to the group generated by $\varphi_\mathfrak{f}(u)$, $u \in J(\mathfrak{f})$, and by roots of unity.

## 3. EISENSTEIN NUMBERS

The bridge between special values of Hecke $L$ series associated with $K$ and elliptic units is provided by Eisenstein series. We use the term "Eisenstein numbers" for their special values at CM points on the modular curve, because their role parallels the role Bernoulli numbers play in the cyclotomic theory. This section owes a great deal to Weil's book [We2], and also to the paper of Goldstein and Schappacher [Go-Sch].

**3.1** Throughout, let $\mathfrak{f}$ be an integral ideal of $K$ with $w_\mathfrak{f} = 1$, relatively prime to the prime $\mathfrak{p}$. Let $E$ be an elliptic curve with complex multiplication by $\mathcal{O}_K$, defined over $F = K(\mathfrak{f})$. We assume that $E$ satisfies (1.4) $\psi_{E/F} = \varphi \circ N_{F/K}$, and that the conductor of $\varphi$ divides $\mathfrak{f}$. Fix also a (generalized) Weierstrass model of $E$ over $F$, with good reduction at the primes above $\mathfrak{p}$. Let $\omega_E$ be the standard differential on that model, and $L$ its period lattice. Recall that $A(\cdot)$ is the cocycle associated to $L$ (or $\omega_E$) as in 1.5.

Letting $\zeta(z, L) = \sigma'(z, L)/\sigma(z, L)$ stand for *Weierstrass' zeta function*, we observe first of all that

$$(1) \qquad E_1(z, L) = \zeta(z, L) - \eta(z, L)$$

is $L$-periodic. From 2.1(7) and the formula for $\eta(z, L)$ preceding it, another expression is derived:

$$(2) \qquad E_1(z, L) = \frac{1}{12}\frac{\partial}{\partial z} \log \theta(z, L) - \frac{1}{2}\,\overline{z}A(L)^{-1}.$$

Following Weil, put $L = \mathbf{Z}\omega_1 + \mathbf{Z}\omega_2$, $Im(\omega_1/\omega_2) > 0$, and treat $z, \overline{z}, \omega_1, \overline{\omega}_1, \omega_2$ and $\overline{\omega}_2$ as six independent variables. Formulas 2.1(4) express $\eta(z, L)$ in terms of these six variables, and for $\zeta(z, L)$ there is the well-known expansion

$$(3) \quad \zeta(z, L) = \frac{1}{z} + \sum_{m,n}{}' \left\{\frac{1}{z - m\omega_1 - n\omega_2} + \frac{1}{m\omega_1 + n\omega_2} + \frac{z}{(m\omega_1 + n\omega_2)^2}\right\}.$$

Let us introduce two differential operators

$$(4) \qquad \partial = -\frac{\partial}{\partial z}, \qquad \mathcal{D} = -A(L)^{-1}\left(\overline{z}\frac{\partial}{\partial z} + \overline{\omega}_1\frac{\partial}{\partial \omega_1} + \overline{\omega}_2\frac{\partial}{\partial \omega_2}\right).$$

It is easy to check that $\mathcal{D}(A(L)) = 0$, so $A(L)$ is a constant with respect to the two operators.

If $0 \leq -j < k$, and $\mathfrak{a}$ is an integral ideal, define

$$(5) \qquad \begin{cases} E_{j,k}(z, L) = \mathcal{D}^{-j}\partial^{k+j-1}E_1(z, L), \\ E_{j,k}(z; L, \mathfrak{a}) = \mathbf{N}\mathfrak{a} \cdot E_{j,k}(z, L) - E_{j,k}(z, \mathfrak{a}^{-1}L), \\ E_k = E_{0,k}. \end{cases}$$

The next two formulas, easily verified from the definition, are the basis for the connection between $E_{j,k}$ and $L$ functions, on one side, and elliptic units, on the other.

$$(6) \qquad E_{j,k}(z,L) = (k-1)!A(L)^j \cdot \sum_{\omega \in L} (z+\omega)^{-k}(\overline{z}+\overline{\omega})^{-j}, \qquad k+j \geq 3,$$

$$(7) \qquad -12 \cdot E_k(z;L,\mathfrak{a}) = \partial^k \log \Theta(z;L,\mathfrak{a}), \qquad k \geq 1.$$

**3.2** A crucial step in the study of the special values of $E_{j,k}(z,L)$ is provided by the lemma below. It will allow us in 4.14 to replace the $E_{j,k}$ by expressions involving $E_k$ only.

**Lemma.** *There exists a unique polynomial $\Phi_{j,k}$ in $\mathbf{Z}[X_1,\ldots,X_{k-j}]$, of degree $1-j$, isobaric of weight $k-j$ ($X_i$ is assigned weight $i$), such that*

$$E_{j,k} = 2^j \Phi_{j,k}(E_1,\ldots,E_{k-j}).$$

*Furthermore,*

$$\Phi_{j,k} = (-2X_1)^{-j}X_k + (\textit{terms in which } X_1 \textit{ appears to degree} < -j).$$

PROOF: The first statement is proved by Weil in [We2], VI §4. The fact that $\Phi_{j,k}$ has integral coefficients and the second assertion are not explicitly mentioned there, but follow easily from the proof.

**3.3** Let us summarize some properties of $E_{j,k}$ that will be needed later. In view of (7), it is not surprising that they look very much like the facts proven about $\Theta(v;L,\mathfrak{a})$ in §2.

Let $\mathfrak{m}$ be any non-trivial integral ideal, and $v$ a primitive $\mathfrak{m}$ division point of $L$. Recall that $L$ was the period lattice of a *good* Weierstrass model (at $\mathfrak{p}$), defined over $K(\mathfrak{f}) = F$, and $0 \leq -j < k$.

**Proposition.** *(i) $E_{j,k}(cz,cL) = c^{j-k}E_{j,k}(z,L)$.*

*(ii) (Rationality). $E_{j,k}(v,L) \in K(l.c.m.\ (\mathfrak{f},\mathfrak{m}))$.*

*(iii) (Galois action). If $\mathfrak{c}$ is integral, $(\mathfrak{c},\mathfrak{f}\mathfrak{m}) = 1$, then*

$$(8) \qquad E_{j,k}(v,L)^{\sigma_\mathfrak{c}} = E_{j,k}(\Lambda(\mathfrak{c})v, \Lambda(\mathfrak{c})\mathfrak{c}^{-1}L) = \Lambda(\mathfrak{c})^{j-k}E_{j,k}(v,\mathfrak{c}^{-1}L).$$

*(iv) (Integrality). Suppose that* $\mathfrak{m}$ *is not a power of* $\mathfrak{p}$*, and that* $\mathfrak{p}$ *is split in* $K/\mathbf{Q}$*. Then* $\overline{\mathfrak{m}}E_1(v,L)$ *and* $(2\overline{\mathfrak{m}})^{-j}E_{j,k}(v,L)$ *for* $1 < k,\ 0 \leq -j < k$*, are* $\mathfrak{p}$*-integral.*

PROOF: (i) This is a consequence of (5).

(ii) Let $\mathfrak{g} = l.c.m.(\mathfrak{f},\mathfrak{m})$. We have already seen (1.6) that $K(\mathfrak{g}) = F(E[\mathfrak{m}])$, where $E$ is the Weierstrass model corresponding to $L$. By the lemma, it is enough to prove (ii) for $E_k,\ k \geq 1$. By (7), $E_k(z;L,\mathfrak{a})$ is an $F$-rational elliptic function, so $E_k(v;L,\mathfrak{a}) \in K(\mathfrak{g})$. Choose $\mathfrak{a} = (\alpha),\ \alpha \equiv 1 \ mod\ \mathfrak{m}$. Then $E_k(v,\mathfrak{a}^{-1}L) = \alpha^k E_k(\alpha v, L) = \alpha^k E_k(v,L)$, because $E_k$ is $L$-periodic. Since we may arrange $\mathbf{N}\alpha \neq \alpha^k$, (5) shows that $E_k(v,L) \in K(\mathfrak{g})$.

(iii) This may be deduced from 2.4(ii) and (7) by the same reasoning as above.

(iv) We shall first prove that $\overline{\mathfrak{m}}E_k(v,L)$ is $\mathfrak{p}$-integral for $k \geq 1$. It is enough to show that when we fix the embedding of $\overline{\mathbf{Q}}$ into $\mathbf{C}_p$ it is a local integer there. We shall prove later on (4.9) that if $(\mathfrak{a},\mathfrak{m}\mathfrak{p}) = 1$, $\Theta(v-z;L,\mathfrak{a})$ has a $\mathfrak{p}$-integral $t$ expansion $G(t)$ at 0. Thus

$$\partial^k \log \Theta(v-z;L,\mathfrak{a}) = \left(-\frac{1}{\lambda'_{\hat{E}}(t)}\frac{d}{dt}\right)^k \log G(t)$$

is also a $\mathfrak{p}$-integral power series. Furthermore, up to a constant, $G(t)$ is a $12^{th}$ power of another power series with $\mathfrak{p}$-integral coefficients. By (7), $E_k(v;L,\mathfrak{a})$ is $\mathfrak{p}$-integral. Choose $\mathfrak{a} = (\alpha),\ \alpha \equiv 1\ mod\ \mathfrak{m}\mathfrak{p}^r$ for some large $r$, such that the same power of $\overline{\mathfrak{p}}$ divides $\alpha - 1$ and $\mathfrak{m}$. Then, if $r$ is large enough, $\overline{\alpha} - \alpha^{k-1}$ and $\overline{\mathfrak{m}}$ are divisible by the same power of $\mathfrak{p}$. Now

$$E_k(v;L,(\alpha)) = \alpha(\overline{\alpha} - \alpha^{k-1}) \cdot E_k(v,L),$$

so $\overline{\mathfrak{m}}E_k(v,L)$ is $\mathfrak{p}$-integral.

To prove the rest of part (iv) we may assume, again by lemma 3.2, that $j = 0$. First observe that with a generalized Weierstrass model 1.11(6'), and with $b_2 = a_1^2 + 4a_2,\ b_4 = a_1a_3 + 2a_4$, as in [Ta], one has

(9) $$\wp'' = 6(\wp - b_2/12)^2 + b_2(\wp - b_2/12) + b_4.$$

Supppose that $k \geq 3$. Then $E_k(v, L) = (-1)^k \wp^{(k-2)}(v, L)$. Since $v$ is a primitive $\mathfrak{m}$-division point and $\mathfrak{m}$ is not a power of $\mathfrak{p}$, $\xi(v, L)$ does not belong to the kernel of reduction, hence its $x$ and $y$ coordinates are integral. It follows from 1.11(19'), (9) and easy induction, that $\wp^{(k-2)}(v, L)$ is integral.

There remains the case $k = 2$. Now, as follows for instance from (1) and 2.1, $E_2(v, L) = \wp(v, L) + s_2(L)$, so by 1.11(19') we only have to show that

$$s_2(L) + \frac{1}{12}(a_1^2 + 4a_2) \tag{10}$$

is a $\mathfrak{p}$-adic integer. To establish this, let $\mathfrak{n}$ be an auxiliary ideal, $(\mathfrak{n}, p) = 1$, and $w$ a primitive $\mathfrak{n}$ torsion point. By the discussion above, $\overline{\mathfrak{n}} E_2(w, L)$, hence $E_2(w, L)$ as well, is integral. Applying the previous argument in reverse,

$$s_2(L) + \frac{1}{12}(a_1^2 + 4a_2) = E_2(w, L) - x(\xi(w, L))$$

is integral. The proof of (iv) is complete.

REMARK: Any change of coordinates in 1.11(6') transforms $b_2 = a_1^2 + 4a_2$ to $b_2' = u^2 b_2 + 12r$ with some unit $u$ and an integer $r$. Choosing $r$ appropriately, we may therefore assume that $s_2(L) = -b_2/12$.

**3.4** CONGRUENCES: The integrality results of proposition 3.3 lead to some very useful congruences between Eisenstein numbers. As in 3.3(iv) *we continue to assume that $p = \mathfrak{p}\overline{\mathfrak{p}}$ is split*, and $i_p : \overline{\mathbf{Q}} \hookrightarrow \mathbf{C}_p$ induces $\mathfrak{p}$ on $K$. We continue to denote by $v$ a primitive $\mathfrak{m}$ division point of $L$. As already implied by the proposition, $E_k(v, L)$ is in general $\mathfrak{p}$-integral only if $k \geq 2$. At places above $\mathfrak{p}$ $E_1(v, L)$ has a denominator whose order is at most, and as we shall see later, precisely, the power of $\overline{\mathfrak{p}}$ in $\mathfrak{m}$.

**Lemma**. *Assume that $\mathfrak{m}$ is not a power of $\mathfrak{p}$. In the following statements, all congruences should be read locally, in $\mathbf{C}_p$.*

*(i)* $E_1(v, L)^{\sigma_\mathfrak{c}} \equiv \mathbf{N}\mathfrak{c}\Lambda(\mathfrak{c})^{-1} E_1(v, L)$ *mod* 1, $(\mathfrak{c}, \mathfrak{f}\mathfrak{m}) = 1$.

*In (ii) and (iii) assume $\mathfrak{f} | \mathfrak{m}$, so that $F(E[\mathfrak{m}]) = K(\mathfrak{m})$.*

*(ii) Let* $\mathfrak{q}$ *be a prime,* $(\mathfrak{q}, \mathfrak{m}\overline{\mathfrak{p}}) = 1$. *Then for any* $u \in \mathfrak{q}^{-1}L$,

$$(\mathrm{N}\mathfrak{q} - 1)E_1(v, L) \equiv (\mathrm{N}\mathfrak{q} - 1)E_1(v + u, L) \ mod\ 1.$$

*(iii) Let* $\mathfrak{g}$ *be integral,* $(\mathfrak{g}, p) = 1$, *and let* $p^r$ *be the power of* $p$ *in* $\prod(\mathrm{N}\mathfrak{q} - 1)$, *the product taken over all the primes* $\mathfrak{q}$ *dividing* $\mathfrak{g}$, *but not* $\mathfrak{m}$. *Then*

$$p^r E_1(v, L) \equiv p^r \mathrm{N}\mathfrak{g} E_1(v, \mathfrak{g}L) \ mod\ 1.$$

REMARK: Parts (ii) and (iii) perhaps hold without the factors $\mathrm{N}\mathfrak{q} - 1$ and $p^r$. However, the weaker congruences recorded here will be sufficient for all our needs in §4. The reader should keep in mind the typical case, where $\mathfrak{m} = \mathfrak{f}\mathfrak{p}^n\overline{\mathfrak{p}}^m$, $n$ fixed, and $m$ very large. It is in this context that the lemma will be used later on.

PROOF: (i) We begin by showing that we may assume $(\mathfrak{c}, \mathfrak{p}) = 1$. If not, let $\alpha \in K^x$ be such that $\mathfrak{b} = (\alpha)\mathfrak{c}$ is integral, $\alpha \equiv 1 \ mod\ \mathfrak{m}\mathfrak{f}\overline{\mathfrak{p}}^n$ for some large $n$, and $(\mathfrak{b}, \mathfrak{p}) = 1$. The left hand side of (i) is unchanged if we replace $\mathfrak{c}$ by $\mathfrak{b}$. Also, $\mathrm{N}\mathfrak{b}\Lambda(\mathfrak{b})^{-1} = \mathrm{N}\mathfrak{c}\Lambda(\mathfrak{c})^{-1}\overline{\alpha}$, so if $n$ is large enough, the right hand side of (i) is unchanged too. Assuming therefore $(\mathfrak{c}, \mathfrak{f}\mathfrak{m}\mathfrak{p}) = 1$, we know that $E_1(v; L, \mathfrak{c}) = \mathrm{N}\mathfrak{c}E_1(v, L) - E_1(v, \mathfrak{c}^{-1}L)$ is $\mathfrak{p}$-integral (see the proof of 3.3(iv)). To derive (i) divide by $\Lambda(\mathfrak{c})$, now a unit, and use (8).

(ii) Observe first that

$$E_1(z, \mathfrak{q}^{-1}L) = \sum_{u \in \mathfrak{q}^{-1}L/L} E_1(z + u, L). \tag{11}$$

This distribution relation may be deduced from proposition 2.3, or more easily, from the fact that $E_1(z, L)$ equals the value at $s = 0$ of the analytic continuation of $\sum (z+\omega)^{-1}|z+\omega|^{-2s}$ ([We2], VIII §14). By 1.7, since $\mathfrak{f}|\mathfrak{m}$, the numbers $E_1(v+u, L)$, for $u \in \mathfrak{q}^{-1}L - L$, are all conjugate under $Gal(K(\mathfrak{m}\mathfrak{q})/K(\mathfrak{m}))$. If $\mathfrak{c}$ is an ideal whose Artin symbol belongs to this Galois group, $\mathrm{N}\mathfrak{c}\Lambda(\mathfrak{c})^{-1} = \overline{\varphi}(\mathfrak{c}) \equiv 1 \ mod\ \overline{\mathfrak{m}}$, so if $u_0 \in \mathfrak{q}^{-1}L - L$,

$$(\mathrm{N}\mathfrak{q} - 1)E_1(v + u_0, L) + E_1(v, L) \equiv \sum_{\mathfrak{c}} \mathrm{N}\mathfrak{c}\Lambda(\mathfrak{c})^{-1}E_1(v + u_0, L) + E_1(v, L)$$

$$\equiv \sum E_1(v + u_0, L)^{\sigma_\mathfrak{c}} + E_1(v, L) = \sum_{u \in \mathfrak{q}^{-1}L/L} E_1(v + u, L) = E_1(v, \mathfrak{q}^{-1}L)$$

$$\equiv \mathrm{N}\mathfrak{q} \cdot E_1(v, L) \ mod\ 1,$$

by part (i) and 3.3(iv).

(iii) An easy induction argument shows that to prove (iii) we may assume that $\mathfrak{g}$ is a prime $\mathfrak{q}$. We distinguish between two cases, according to whether $\mathfrak{q}$ divides $\mathfrak{m}$, or not. In the former case, (11) gives

$$E_1(v,L) = \sum_{u \in L/\mathfrak{q}L} E_1(v+u, \mathfrak{q}L) \equiv \mathbf{N}\mathfrak{q} \cdot E_1(v, \mathfrak{q}L) \; mod \; 1$$

by the same argument as above, since now all $\mathbf{N}\mathfrak{q}$ points $\xi(v+u, \mathfrak{q}L)$ are conjugate under $Gal(K(\mathfrak{m}\mathfrak{q})/K(\mathfrak{m}))$. If $\mathfrak{q} \nmid \mathfrak{m}$, on the other hand,

$$(\mathbf{N}\mathfrak{q} - 1)E_1(v,L) = (\mathbf{N}\mathfrak{q} - 1) \cdot \sum_{u \in L/\mathfrak{q}L} E_1(v+u, \mathfrak{q}L)$$
$$\equiv (\mathbf{N}\mathfrak{q} - 1)\mathbf{N}\mathfrak{q} \cdot E_1(v, \mathfrak{q}L) \; mod \; 1$$

by part (ii). This concludes the proof of the lemma.

For future reference, we mention here another (local) congruence that follows from lemma 3.2 and 3.3(iv). Let $\mathfrak{m}$ be divisible by some prime different from $\mathfrak{p}$, and let $v$ be an $\mathfrak{m}$-primitive division point of $L$, as before. Then, for $0 \leq -j < k$, if $\overline{\mathfrak{m}}|c$,

$$(2c)^{-j} E_{j,k}(v,L) \equiv (-2cE_1(v,L))^{-j} E_k(v,L) \; mod \; \overline{\mathfrak{m}}. \tag{12}$$

**3.5** The results obtained so far may be of some independent interest, but their significance stems mainly from the relation between the numbers $E_{j,k}(v,L)$ and special values of $L$ functions. Recall that the partial $L$ function $L\left(\chi, s; \left(\frac{M/K}{\mathfrak{c}}\right)\right)$ is defined to be $\sum \chi(\mathfrak{a})\mathbf{N}\mathfrak{a}^{-s}$, the sum extending over all integral ideals $\mathfrak{a}$ such that $(\mathfrak{a}, \mathfrak{f}_{M/K}) = 1$, and $(\mathfrak{a}, M/K) = (\mathfrak{c}, M/K)$.

**Proposition.** *Let $\varphi$ be a grossencharacter of type (1,0) whose conductor divides $\mathfrak{m}$. Then for any integral ideal $\mathfrak{c}$, $(\mathfrak{c},\mathfrak{m}) = 1$, and $\Omega \in \mathbf{C}^x$,*

$$\mathbf{N}\mathfrak{m}^{-j} E_{j,k}(\Omega, \mathfrak{c}^{-1}\mathfrak{m}\Omega) = (k-1)! \left(\frac{\sqrt{d_K}}{2\pi}\right)^j \Omega^{j-k} \cdot \varphi(\mathfrak{c})^{k-j} \cdot L\left(\overline{\varphi}^{k-j}, k; \left(\frac{K(\mathfrak{m})/K}{\mathfrak{c}}\right)\right). \tag{13}$$

PROOF: This is a straightforward computation, which, at least when $k + j \geq 3$, boils down to (6). In the remaining cases one has to use Hecke's trick of introducing $|z+\omega|^{-2s}$ into (6) and analytically continuing in $s$. The relevant fact here is that $E_{j,k}$ is the value of this analytic continuation at $s = 0$ ([We2], VIII §14). See also [Go-Sch], corollary 5.7.

Note that if $\Omega$ is chosen so that $L = \mathfrak{m}\Omega$ is the lattice of a Weierstrass model $E$ as in 3.1, the left hand side of (13) belongs to $K(\mathfrak{m})$, hence so does the right hand side. This was originally discovered by Damerell [Da]. In this case $E_{j,k}(\Omega, \mathfrak{c}^{-1}\mathfrak{m}\Omega) = \Lambda(\mathfrak{c})^{k-j} E_{j,k}(\Omega, \mathfrak{m}\Omega)^{\sigma_\mathfrak{c}}$.

## 4. $p$-ADIC $L$ FUNCTIONS

We have now developed all the ingredients needed to construct the $p$-adic $L$ functions in full generality. Chapter I 3.1-3.6 contains a recipe for turning a norm-coherent system of local units into a measure. Section 2 of this chapter supplies interesting examples of such norm-coherent sequences, namely the elliptic units. Section 3 relates, as we shall see below, their Coleman power series to special values of classical $L$ functions. We now combine everything.

From now on $p$ will denote a prime that splits in $K$

$$p = \mathfrak{p}\overline{\mathfrak{p}} \tag{1}$$

and we assume that $\mathfrak{p}$ is the prime induced from the fixed embedding of $\overline{\mathbf{Q}}$ in $\mathbf{C}_p$. The splitting restriction is imposed upon us because of the fundamental assumption made in I.3.2, that the formal group is *of height* 1.

Traditionally, the $p$-adic $L$ functions were labeled either *one-variable* or *two-variables.* Since our construction employs measures rather than power series, this distinction is not so prominent. See 4.16 for comparison with power series terminology.

The main theorems are formulated in 4.12 and 4.14, and the reader is advised to look them up before he is deluged with computations. It is perhaps worth noting from the outset, that to understand the statements alone, nothing about elliptic curves is needed. The *construction* requires the elliptic curves studied in 1.4, whose torsion is abelian over $K$.

**4.1** Fix an integral ideal $\mathfrak{f}$ of $K$, with $w_{\mathfrak{f}} = 1$, and relatively prime to $\mathfrak{p}$. Let $F = K(\mathfrak{f})$ and $F_n = K(\mathfrak{f}\mathfrak{p}^n)$. We emphasize that $\mathfrak{f}$ need not be relatively prime to $\overline{\mathfrak{p}}$. In fact, our chief interest in 4.14 will be in what happens when one lets $\mathfrak{f}\overline{\mathfrak{p}}^m$ stand for the original $\mathfrak{f}$.

Fix a grossencharacter $\varphi$ of type (1,0), whose conductor divides $\mathfrak{f}$, and an elliptic curve $E$ over $F$, satisfying 1.4 (12), for which $\psi_{E/F} = \varphi \circ N_{F/K}$. Such $\varphi$ and $E$ exist by lemma 1.4. For any $\mathfrak{g}$ divisible by $\mathfrak{f}$, $K(\mathfrak{g}) = F(E[\mathfrak{g}])$, and this in particular holds with $\mathfrak{g} = \mathfrak{f}\mathfrak{p}^n$. Furthermore, $Gal(F_n/F) \cong (\mathcal{O}_K/\mathfrak{p}^n)^\times$ via its action on $E[\mathfrak{p}^n]$ (corollary 1.7).

It is easy, nevertheless essential to what follows, to give a *semi-local* version of the local results from chapter I, considering all places above $\mathfrak{p}$ simultaneously. This we shall now carry out. Let us put

$$(2) \qquad \Phi = F \otimes_K K_{\mathfrak{p}} = \oplus_{\mathfrak{P}|\mathfrak{p}} F_{\mathfrak{P}}, \qquad R = \mathcal{O}_F \otimes_{\mathcal{O}_K} \mathcal{O}_{\mathfrak{p}} = \oplus_{\mathfrak{P}|\mathfrak{p}} \mathcal{O}_{\mathfrak{P}},$$

and let $\Phi_n$ and $R_n$ be defined similarly, with $F_n$ instead of $F$. Recall that above each prime $\mathfrak{P}$ of $F$ dividing $\mathfrak{p}$, lies a unique prime $\mathfrak{P}_n$ of $F_n$.

Let $\mathcal{G} = Gal(F_\infty/K)$, $G = Gal(F_\infty/F) = \Gamma \times \Delta$, where $\Gamma \cong 1 + p\mathbf{Z}_p$, $\Delta \cong \mathbf{F}_p^\times$. Needless to say, $\mathcal{G}$ acts on $\Phi_n$ via its action on $F_n$.

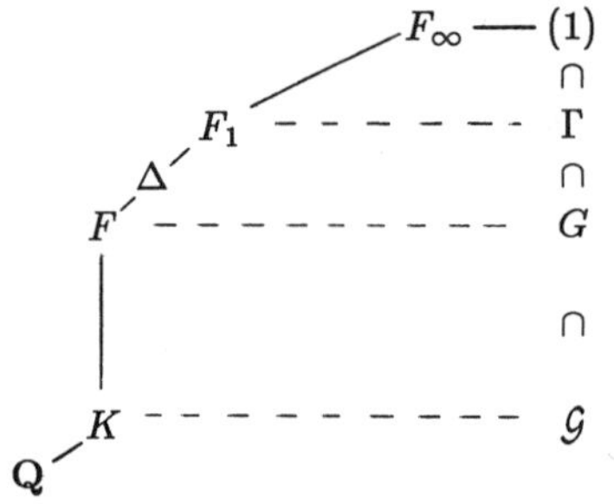

Let $R_n^x = U_n \times V_n$ be the decomposition of the semi-local units into a pro-$p$ part $U_n$ (principal units), and a finite group $V_n$ of order prime to $p$. Obviously, $V_n$ is independent of $n$. If $[F : K] = fg$, and $\mathfrak{p}$ decomposes into $g$ $\mathfrak{P}$'s, each of norm $q = p^f$, then $V_n \cong (\mathbf{F}_q^x)^g$.

Finally define inverse systems with respect to the norm map

$$\mathcal{U} = \varprojlim U_n, \qquad \mathcal{V} = \varprojlim V_n. \tag{3}$$

**4.2** From now on let $E$ stand for a particular Weierstrass model defined over the localization of $\mathcal{O}_F$ at $\mathfrak{p}$, with *good reduction* above $\mathfrak{p}$, i.e. $\Delta$ is a unit at each place $\mathfrak{P}$ dividing $\mathfrak{p}$. The models $E^\sigma$, $\sigma \in Gal(F/K)$, are of the same type, and what is said here about $E$, applies equally well to them.

The formal group $\hat{E}$, with its parameter $t$ corresponding to the specific Weierstrass model fixed above, is defined over $R$. At each $\mathfrak{P}$, $\hat{E}$ projects to a Lubin-Tate formal group of height 1, relative to the unramified extension $F_{\mathfrak{P}}/K_{\mathfrak{p}}$, which is defined over $\mathcal{O}_{\mathfrak{P}}$. We denote it by $\hat{E}_{\mathfrak{P}}$ (1.10).

If $(\mathfrak{a}, \mathfrak{f}) = 1$, the isogeny $\lambda(\mathfrak{a})$ induces a homomorphism of formal groups over $R$

$$\widehat{\lambda(\mathfrak{a})} : \hat{E} \to \widehat{E^{\sigma_\mathfrak{a}}}. \tag{4}$$

If $\sigma = \sigma_\mathfrak{a}$ belongs to the decomposition group of $\mathfrak{p}$ in $Gal(F/K)$, this homomorphism breaks down as the product of $\hat{E}_{\mathfrak{P}} \to \hat{E}_{\mathfrak{P}}^\sigma$ for $\mathfrak{P}|\mathfrak{p}$. In the notation of I.1 these arrows are $[\Lambda(\mathfrak{a})]_{f,\sigma(f)}$, and indeed $\Lambda(\mathfrak{a}) \in F_{\mathfrak{P}}$ satisfies $\Lambda(\mathfrak{a})^{\phi-1} = \Lambda(\mathfrak{p})^{\sigma-1}$ ($\phi = \sigma_\mathfrak{p}$), which must hold in order for $[\Lambda(\mathfrak{a})]_{f,\sigma(f)}$ to make sense (proposition I.1.5). Furthermore, if $(\mathfrak{a}, \mathfrak{f}\mathfrak{p}) = 1$ and $(\mathfrak{a}, F/K) = 1$, then $\widehat{\lambda(\mathfrak{a})} \in End(\hat{E})$, and comparing 1.5 (15) to I.3.3 (9), we obtain

$$\Lambda(\mathfrak{a}) = \kappa(\sigma_\mathfrak{a}). \tag{5}$$

With the convention of I.3.7 (14), regarding the extension of $\kappa$ from $G$ to $\mathcal{G}$, (5) remains valid even without the restriction $(\mathfrak{a}, F/K) = 1$. Now, of course, $\Lambda(\mathfrak{a}) \in F$, and $\kappa(\sigma_\mathfrak{a}) \in \Phi$.

We let $L$, as before, be the period lattice of $E$, and $L_{\mathfrak{a}} = \Lambda(\mathfrak{a})\mathfrak{a}^{-1}L$ that of $E^{\sigma_{\mathfrak{a}}}$. Replacing $E$ by one of its conjugates, if necessary, we assume

$$(6) \qquad L = \Omega\mathfrak{f}, \qquad \Omega \in \mathbf{C}^{x}.$$

Fix a choice of $\Omega$ once and for all, and note that $\varphi$ and (6) determine $E$ up to $F$-isomorphism, so $\Omega$ *mod* $F^{x}$ is independent of the specific Weierstrass model.

**4.3** $p$-ADIC PERIODS: Let $F' = F(E[\overline{\mathfrak{p}}^{\infty}])$, and

$$(7) \qquad \Phi' = F' \otimes_K K_{\mathfrak{p}} \qquad R' = \mathcal{O}_{F'} \otimes_{\mathcal{O}_K} \mathcal{O}_{\mathfrak{p}}.$$

Since $\mathfrak{p}$ is finitely decomposed and unramified in $F'$, $\Phi'$ is a finite direct sum of fields, each unramified over $K_{\mathfrak{p}}$. Let $\hat{\Phi}$ and $\hat{R}$ be the completions of $\Phi'$ and $R'$. Clearly $Gal(F'/K)$ acts on $\Phi'$ continuously, via its action on $F'$, and the action extends to $\hat{\Phi}$.

**Proposition.** *There exists an isomorphism of formal groups defined over* $\hat{R}$,

$$(8) \qquad \theta : \hat{\mathbf{G}}_m \simeq \hat{E}, \qquad t = \theta(S) = \Omega_p S + \cdots \in \hat{R}[[S]],$$

*satisfying*

$$(9) \qquad \widehat{\lambda(\mathfrak{c})} \circ \theta = \theta^{\sigma_{\mathfrak{c}}} \circ [\mathbf{N}\mathfrak{c}]_{\hat{\mathbf{G}}_m}, \qquad (\mathfrak{c}, \mathfrak{f}\overline{\mathfrak{p}}) = 1.$$

$\Omega_p \in \hat{R}^{x}$ *is uniquely determined by (9) modulo* $\mathcal{O}_{\mathfrak{p}}^{x}$.

PROOF: This is the semi-local version of I.1.6 and I.3.2 (3). First observe that by Tate's theorem on invariants ([Ta2], p. 176) the fixed subring of $\hat{\Phi}$ under $Gal(F'/K)$ is $K_{\mathfrak{p}}$. Since (9) implies

$$(10) \qquad \Omega_p^{\sigma_{\mathfrak{c}}-1} = \Lambda(\mathfrak{c})\mathbf{N}\mathfrak{c}^{-1}, \qquad (\mathfrak{c}, \mathfrak{f}\overline{\mathfrak{p}}) = 1,$$

and these $\sigma_{\mathfrak{c}}$ are dense in $Gal(F'/K)$, the last assertion follows.

To construct $\theta$ we first find an $\Omega_p$ as in (10). If $\sigma_{\mathfrak{c}}$ fixes $F$ and $E[\overline{\mathfrak{p}}^{m}]$ pointwise, then (1.5 (18)) $\Lambda(\mathfrak{c}) = \varphi(\mathfrak{c})$, and by Weil's pairing, or from $\varphi(\mathfrak{c})\overline{\varphi(\mathfrak{c})} = \mathbf{N}\mathfrak{c}$,

$\Lambda(\mathfrak{c})\mathbf{N}\mathfrak{c}^{-1} \equiv 1 \; mod \; \mathfrak{p}^m$. It follows that the map $\sigma_\mathfrak{c} \mapsto \Lambda(\mathfrak{c})\mathbf{N}\mathfrak{c}^{-1}$ extends to a continuous 1-cocycle $Gal(F'/K) \to R^x \subset \hat{R}^x$ (see 1.5 (17)). Since $F'/K$ is unramified at $\mathfrak{p}$, Hilbert's theorem 90 implies $H^1(Gal(F'/K), \hat{R}^x) = 1$, so $\Omega_p$ exists (compare I.1.6).

Proceed as in chapter I. By the semi-local analogue of lemma I.1.4, there exists a power series $\theta$ with $\widehat{\lambda(\mathfrak{p})} \circ \theta = \theta^{\sigma_\mathfrak{p}} \circ [p]_{\hat{\mathbf{G}}_m}$, $\theta(S) = \Omega_p S + \cdots$, and this implies that $\theta : \hat{\mathbf{G}}_m \simeq \hat{E}$. Finally (9) holds because both sides are homomorphisms from $\hat{\mathbf{G}}_m$ to $\hat{E}^{\sigma_\mathfrak{c}}$ with the same derivative at 0.

**4.4** CHOOSING $\Omega_p$: Fix once and for all a generator $(\varsigma_n)$ of the Tate module of $\hat{\mathbf{G}}_m$, as in I.3.2. Such a choice should be regarded as *orienting* $\mathbf{C}_p$. Now, it is clear from proposition 4.3 that the following are equivalent:

(i) a choice of $\Omega_p$ as in (10),

(ii) a choice of $\theta$ as in (8).

Let

$$\omega_n = \theta^{\phi^{-n}}(\varsigma_n - 1) \in R_n \tag{11}$$

(see I.3 (5), $\phi = \sigma_\mathfrak{p}$). Then there exists a unique point $u_n \in L/\mathfrak{p}^n L$ such that

$$\omega_n = t(\xi(\Lambda(\mathfrak{p}^{-n})u_n, \Lambda(\mathfrak{p}^{-n})\mathfrak{p}^n L)). \tag{12}$$

Recall that $\xi(\cdot, \Lambda(\mathfrak{p}^{-n})\mathfrak{p}^n L) : \mathbf{C}/\Lambda(\mathfrak{p}^{-n})\mathfrak{p}^n L \simeq E^{\phi^{-n}}(\mathbf{C})$. Furthermore, since $(\phi^{-n}\widehat{\lambda(\mathfrak{p})})(\omega_n) = \omega_{n-1}$, $u_n \; mod \; \mathfrak{p}^{n-1}L = u_{n-1}$. See 1.5 (21). It is now clear that (i) and (ii) are also equivalent to either of the following:

(iii) a choice of $(\omega_n)$, for which $\phi^{-n}\widehat{\lambda(\mathfrak{p})}(\omega_n) = \omega_{n-1}$, and $\phi^{-n}\widehat{\lambda(\mathfrak{c})}(\omega_n) = \sigma_\mathfrak{c}(\omega_n)$, $(\mathfrak{c}, \mathfrak{f}\mathfrak{p}) = 1$,

(iv) a choice of $u_n \in L/\mathfrak{p}^n L$, primitive of level $\mathfrak{p}^n$, such that $u_n \equiv u_{n-1} \; mod \; \mathfrak{p}^{n-1}L$, for each $n \geq 1$.

We are now going to show how, with a fixed orientation of $\mathbf{C}_p$, $\Omega$ determines $\Omega_p$ *canonically*. Let $w_n$ be the unique $\mathfrak{f}$-division point of $\mathfrak{p}^n L$ for which $\phi^n(\xi(\Lambda(\mathfrak{p}^{-n})w_n, \Lambda(\mathfrak{p}^{-n})\mathfrak{p}^n L)) = \xi(\Omega, L)$. Diagram 1.5 (21) implies that

$w_n \equiv w_{n-1}$ *mod* $\mathfrak{p}^{n-1}L$, and in particular $w_n \equiv w_0 \equiv \Omega$ *mod* $L$. If we define $u_n = w_n - \Omega$ *mod* $\mathfrak{p}^n L$, we obtain a sequence $(u_n)$ as in (iv).

DEFINITION. *With* $L = \mathfrak{f}\Omega$ *(6), let* $w_n$ *and* $u_n$ *be the unique* $\mathfrak{f}$*- and* $\mathfrak{p}^n$*-division points of* $\mathfrak{p}^n L$ *for which* $w_n - u_n \equiv \Omega$ *mod* $\mathfrak{p}^n L$. *Let* $\omega_n$ *be determined by (12),* $\theta$ *by (11), and* $\Omega_p$ *(the* $p$ *– adic period corresponding to* $\Omega$*) by (8).*

It is easy to verify that $< \Omega, \Omega_p > \in (\mathbf{C}^x \times \mathbf{C}_p^x)/\overline{\mathbf{Q}}^x$ is independent of $\Omega$; it only depends on the fixed choice of $(\varsigma_n)$. See also remark (iv) to theorem 4.11. In fact, more is true; the pair $< \Omega, \Omega_p >$ is well defined modulo $F^x$, and even modulo those elements of $F^x$ which are units at $\mathfrak{p}$.

**4.5** The following proposition is the semi-local version of I.2.2.

**Proposition.** *Let* $\beta = (\beta_n) \in \mathcal{U}$. *There exists a unique power series* $g_\beta(T) \in R[[T]]^x$ *for which*

$$\beta_n = (\phi^{-n} g_\beta)(\omega_n), \qquad n \geq 1. \tag{13}$$

*Furthermore the following properties hold:*

*(i)* $\beta_0 = g_\beta(0)^{1-\phi^{-1}}$,

*(ii)* $g_{\beta\beta'} = g_\beta \cdot g_{\beta'}$,

*(iii)* $g_\beta^\phi \circ \widehat{\lambda(\mathfrak{p})}(T) = \prod_{\omega \in \hat{E}[\mathfrak{p}]} g_\beta(T[+]\omega)$,

*(iv)* $g_{\sigma_\mathfrak{c}(\beta)} = \sigma_\mathfrak{c}(g_\beta) \circ \widehat{\lambda(\mathfrak{c})}, \qquad (\mathfrak{c}, \mathfrak{f}\mathfrak{p}) = 1.$

PROOF: See chapter I, 2.2 and 2.3. For point (iv) see also I.3.7, especially (15).

**4.6** Similarly, the following proposition summarizes the results of I.3.2-3.4 in a semi-local framework.

**Proposition.** *There exists a unique* $\mathcal{G}$ *– homomorphism* $i : \mathcal{U} \rightarrow \Lambda(\mathcal{G}, \hat{R})$ *(=* $\hat{R}$ *– valued measures on* $\mathcal{G}$*)* $i(\beta) = \mu_\beta$, *satisfying*

$$\widetilde{\log g_\beta} \circ \theta(S) = \int_G (1 + S)^{\kappa(\sigma)} \, d\mu_\beta(\sigma), \tag{14}$$

*where* $\kappa : G \simeq \mathbf{Z}_p^x$ *is the character giving the action on* $E[\mathfrak{p}^\infty]$, *and where* $\widetilde{\log g}$ *is defined by I.3.3 (7), with* $W_f^1 = \hat{E}[\mathfrak{p}]$. *The measure* $\mu_\beta$ *depends on the choice of* $(\varsigma_n)$, *but not on* $\theta$ *(or equivalent data as in 4.4).*

**4.7** Recall that we have fixed an embedding of $\overline{\mathbf{Q}}$ into $\mathbf{C}_p$. This means that $\hat{\Phi}$ is mapped into $\mathbf{C}_p$ in a way that sends one of its field components isomorphically onto a subfield of $\mathbf{C}_p$, and the rest to 0. The image of $\mu_\beta$ under this map will be denoted $\mu_\beta^0$; it is an integral $p$-adic measure on $\mathcal{G}$. The reason for introducing $\mu_\beta^0$ in lieu of $\mu_\beta$ is that we wish to integrate certain $\mathbf{C}_p$-valued characters of $\mathcal{G}$ against it, and these characters are *not* $\hat{\Phi}$-valued in general. The next two lemmas complement I.3.5 and I.3.6.

**Lemma.** *Let $\chi$ be a character of $Gal(F/K)$, and $\varphi$ the grossencharacter of type (1,0) fixed in 4.1. For $k \geq 0$ and $\beta \in \mathcal{U}$ let*

$$(15) \qquad \delta_k(\beta) = D^k \log(g_\beta \circ \theta)(0), \qquad D = (1+S)\frac{d}{dS},$$

*be Kummer's logarithmic derivatives (I.3.5), and let $\delta_k(\beta)^0$ be their projection from $\hat{R}$ to $\mathbf{C}_p$. Choose ideals $\mathfrak{c}$ of $K$, relatively prime to $\mathfrak{fp}$, whose Artin symbols $(\mathfrak{c}, F/K)$ represent $Gal(F/K)$. Then, for any $k \geq 0$ the following equality holds:*

$$(16) \qquad \left(1 - \frac{\chi\varphi^k(\mathfrak{p})}{p}\right) \cdot \sum_{\mathfrak{c}} \chi\varphi^k(\mathfrak{c}^{-1}) \cdot \delta_k(\sigma_\mathfrak{c}(\beta))^0 = \int_{\mathcal{G}} \chi\varphi^k(\sigma) d\mu_\beta^0(\sigma).$$

*Here we regard $\varphi$ also as a p-adic character of $\mathcal{G}$ (cf. remark at the end of 1.1). In particular $\varphi(\sigma) = \kappa(\sigma)$ for $\sigma$ in $G$.*

PROOF: If $\tilde{\delta}_k(\beta)$ is defined as in (15), but with $\widetilde{\log}(g_\beta \circ \theta)$ (I.3.3 (7')), then

$$(17) \qquad \tilde{\delta}_k(\beta) = \delta_k(\beta) - p^{k-1}\sigma_\mathfrak{p}(\delta_k(\beta)).$$

Also, if $(\mathfrak{c}, \mathfrak{fp}) = 1$, it follows from proposition 4.5(iv) and from (9), that

$$(18) \qquad \delta_k(\sigma_\mathfrak{c}(\beta)) = \mathbf{N}\mathfrak{c}^k \cdot \sigma_\mathfrak{c}(\delta_k(\beta)).$$

Substituting in the left hand side of (16),

$$\left(1 - \frac{\chi\varphi^k(\mathfrak{p})}{p}\right)\sum_{\mathfrak{c}} \chi\varphi^k(\mathfrak{c}^{-1}) \cdot \mathbf{N}\mathfrak{c}^k \cdot \sigma_{\mathfrak{c}}(\delta_k(\beta))^0$$

$$= \sum_{\mathfrak{c}} \chi\varphi^k(\mathfrak{c}^{-1}) \cdot \tilde{\delta}_k(\sigma_{\mathfrak{c}}(\beta))^0 \qquad \text{(from (17) and (18))}$$

$$= \sum_{\mathfrak{c}} \chi\varphi^k(\mathfrak{c}^{-1}) \cdot \int_G \varphi(\sigma)^k \, d\mu_{\sigma_{\mathfrak{c}}(\beta)}(\sigma) \qquad \text{(I.3.5 (11))}$$

$$= \sum_{\mathfrak{c}} \chi\varphi^k(\mathfrak{c}^{-1}) \cdot \int_G \varphi(\sigma)^k d\mu_\beta^0(\sigma\sigma_{\mathfrak{c}}^{-1})$$

$$= \int_{\mathcal{G}} \chi\varphi^k(\sigma) d\mu_\beta^0(\sigma).$$

**4.8 Lemma.** *Let $\chi$ be a character of $Gal(F_n/K)$, $n \geq 1$, and suppose that $n$ is the exact power of $\mathfrak{p}$ in its conductor. Define the Gauss sum*

$$(19) \qquad \tau(\chi) = \frac{1}{p^n} \sum_{\gamma \in Gal(F_n/F)} \chi(\gamma)\varsigma_n^{-\kappa(\gamma)}$$

*(this is well defined because $\gamma$ determines $\kappa(\gamma)$ modulo $p^n$). For $k \geq 0$ and $\beta \in \mathcal{U}$ let*

$$(20) \qquad \delta_{k,n}(\beta) = D^k \log(g_\beta \circ \theta)(\varsigma_n - 1), \qquad D = (1 + S)\frac{d}{dS}$$

*(in $\hat{\Phi} \otimes K_{\mathfrak{p}}(\varsigma_n)$), and let $\delta_{k,n}(\beta)^0$ be its projection to $\mathbf{C}_p$. Choose ideals $\mathfrak{c}$, relatively prime to $\mathfrak{f}\mathfrak{p}$, whose Artin symbols $(\mathfrak{c}, F_n/K)$ represent $Gal(F_n/K)$. Then*

$$(21) \qquad \tau(\chi) \cdot \sum_{\mathfrak{c}} \chi\varphi^k(\mathfrak{c}^{-1}) \cdot \delta_{k,n}(\sigma_{\mathfrak{c}}(\beta))^0 = \int_{\mathcal{G}} \chi\varphi^k(\sigma) d\mu_\beta^0(\sigma).$$

PROOF: This time we begin the computation on the right hand side. Just as in I.3.6, it is easy to see that

$$\int_{\mathcal{G}} \chi\varphi^k(\sigma) d\mu_\beta^0(\sigma)$$

$$= \sum_{\mathfrak{c}} \chi\varphi^k(\mathfrak{c}^{-1}) \cdot \int_{G_n} \varphi^k(\sigma) d\mu^0_{\sigma_{\mathfrak{c}}(\beta)}(\sigma) \qquad G_n = Gal(F_\infty/F_n)$$

$$= \sum_{\mathfrak{c}} \chi\varphi^k(\mathfrak{c}^{-1}) \cdot \frac{1}{p^n} \sum_{j=0}^{p^n-1} D^k \log(g_{\sigma_{\mathfrak{c}}(\beta)} \circ \theta)(\varsigma_n^j - 1) \cdot \varsigma_n^{-j}.$$

If $\gamma \in G$ and $\kappa(\gamma) \equiv j \bmod p^n$, then

$$(22) \qquad D^k \log(g_{\sigma_{\mathfrak{c}}(\beta)} \circ \theta)(\zeta_n^j - 1) = \kappa(\gamma)^{-k} \cdot D^k \log(g_{\sigma_{\mathfrak{c}}\gamma(\beta)} \circ \theta)(\zeta_n - 1),$$

as follows from proposition 4.5(iv). Now for each $\mathfrak{c}$, and each $j$ relatively prime to $p$, there exists a unique $\gamma \in G$ as above such that $\sigma_{\mathfrak{c}}\gamma = \sigma_{\mathfrak{c}'}$ is again one of the representatives chosen for $Gal(F_n/K)$. It follows that the part of the double sum corresponding to $(j, p) = 1$, is equal to the left hand side of (21). On the other hand, the terms with $p|j$ disappear after summing over the $\mathfrak{c}$'s, because $n$ is the *exact* power of $\mathfrak{p}$ in $\mathfrak{f}_\chi$. To see this, let $\mathfrak{b}$ be such that $\sigma_{\mathfrak{b}} \in Gal(F_\infty/F_{n-1})$. Then

$$D^k \log(g_{\sigma_{\mathfrak{b}}(\beta)} \circ \theta)(\zeta_{n-1}^a - 1) = \varphi^k(\mathfrak{b}) \cdot D^k \log(g_\beta \circ \theta)(\zeta_{n-1}^a - 1)$$

(a special case of (22)). If $j = pa$, break the collection of $\mathfrak{c}$ into cosets modulo $Gal(F_\infty/F_{n-1})$, and use this last formula together with

$$\sum_{\sigma \in Gal(F_n/F_{n-1})} \chi(\sigma) = 0$$

to verify our claim.

**4.9** Lemmas 4.7 and 4.8 (which easily admit a unified formulation) allow us to compute the integrals of $\chi\varphi^k$ against $\mu_\beta$, if we only know $g_\beta$ explicitly. It is time to bring forth the elliptic units.

Let $\mathfrak{a}$ be an integral ideal, relatively prime to $\mathfrak{f}\mathfrak{p}$, and define, for each $n \geq 0$,

$$(23) \qquad e_n(\mathfrak{a}) = \Theta(\Omega; \mathfrak{p}^n L, \mathfrak{a}).$$

*Then $e_n(\mathfrak{a})$, $n \geq 1$, is a unit in $F_n$ (prop. 2.4) and $N_{m,n}e_m(\mathfrak{a}) = e_n(\mathfrak{a})$ for $m \geq n \geq 1$ (prop. 2.3; see also the first proof of prop. 2.4(iii)).* We let

$$(24) \qquad e(\mathfrak{a}) = \varprojlim e_n(\mathfrak{a}) \qquad (w.r.t.\ N_{m,n})$$

and denote by $\beta(\mathfrak{a})$ the projection of $e(\mathfrak{a})$ to the pro-$p$ part $\mathcal{U}$ of $\varprojlim R_n^x$. Our first task is to compute the Coleman power series $g_{e(\mathfrak{a})}$.

Let $\lambda_{\hat{E}} \in \Phi[[T]]$ be the logarithm of the formal group $\hat{E}$, normalized to $\lambda'_{\hat{E}}(0) = 1$.

**Proposition.** *Let $P(z) \in F[[z]]$ be the Taylor series expansion of $\Theta(\Omega - z; L, \mathfrak{a})$. Let $Q(T) = P(\lambda_{\hat{E}}(T))$ (operating formally with power series in $\Phi[[T]]$). Then*

*(i)* $Q(T) \in R[[T]]^{x}$,

*(ii)* $e_n(\mathfrak{a}) = (\phi^{-n}Q)(\omega_n), \qquad n \geq 0.$

*Thus $g_{e(\mathfrak{a})} = Q(T)$, and $g_{\beta(\mathfrak{a})}/g_{e(\mathfrak{a})}$ is a constant.*

PROOF: First observe that since $\omega_E = dz$ is an $F$-rational differential, $z$ is an $F$-rational local parameter on $E$ at the origin. Since $E[\mathfrak{f}] \subset F$, the elliptic function $\Theta(\Omega - z; L, \mathfrak{a})$ is defined over $F$, hence $P(z)$ indeed belongs to $F[[z]]$. This allows us to move from the complex domain to the $p$-adics, and $Q(T) \in \Phi[[T]]$ is well defined.

We claim that as local parameters at the origin, $z$ and $t$ are related via $z = \lambda_{\hat{E}}(t)$. Indeed, since the formal group law in the $z$ variable is the additive law, $z$ is *some* logarithm of $t$. Since $dt/dz(0) = 1$, $z$ is the normalized logarithm of $t$. Thus $Q(T)$ is nothing but the expansion of $\Theta(\Omega - z; L, \mathfrak{a})$ in terms of $t$ at 0.

To verify (i) one can argue by "pure thought" that the function $\Theta(\Omega - z; L, \mathfrak{a})$ has good reduction at $\mathfrak{p}$, as may be read off 2.3 (10). But it is instructive to compute its $t$-expansions, using [Ta] (14) (caution: Tate's $z$ is our $t$), and the addition law on a cubic. In a generalized Weierstrass form 1.11 (6'), the sum of two points $P_1 = (x_1, y_1)$ and $P_2 = (x_2, y_2)$ has $x$-coordinate

$$\left(\frac{y_2 - y_1}{x_2 - x_1}\right)^2 + a_1\left(\frac{y_2 - y_1}{x_2 - x_1}\right) - a_2 - x_1 - x_2.$$

Now $\Delta(L)$ and $\Delta(\mathfrak{a}^{-1}L) \in R^{x}$, so it is enough to find the $t$-expansion of $\wp(\Omega - z, L) - \wp(v, L)$ for $v \in \mathfrak{a}^{-1}L/L - \{0\}$. The formula above shows that this is a power series in $t$, with $\mathfrak{p}$-integral coefficients and constant term $\wp(\Omega, L) - \wp(v, L)$. But this is a $\mathfrak{p}$-adic unit since $(\mathfrak{f}, \mathfrak{a}) = 1$ and both ideals are relatively prime to $\mathfrak{p}$ (see the second proof of 2.4(iii) for similar arguments).

To prove (ii) we must use the choice of $(\omega_n)$ made in 4.4. Recall that $w_n$ was defined to be the unique $\mathfrak{f}$-division point of $\mathfrak{p}^n L$ for which

$$\phi^n(\xi(\Lambda(\mathfrak{p}^{-n})w_n, \Lambda(\mathfrak{p}^{-n})\mathfrak{p}^n L)) = \xi(\Omega, L). \tag{25}$$

It follows that $(\phi^{-n}P)(z) = \Theta(\Lambda(\mathfrak{p}^{-n})w_n - z; \Lambda(\mathfrak{p}^{-n})\mathfrak{p}^n L, \mathfrak{a})$. In the $t$-expansion of this function (on the elliptic curve $E^{\phi^{-n}}$) substitute $t = \omega_n$. In view of 4.4 (12) we obtain the value

$$\Theta(\Lambda(\mathfrak{p}^{-n})w_n - \Lambda(\mathfrak{p}^{-n})u_n; \Lambda(\mathfrak{p}^{-n})\mathfrak{p}^n L, \mathfrak{a}),$$

which is precisely $e_n(\mathfrak{a})$, thanks to the choice of $u_n$. This concludes the proof of the proposition.

**4.10** To get an idea what $\mu_{\beta(\mathfrak{a})}$ looks like, lemmas 4.7 and 4.8 tell us, we have to compute $\delta_{k,n}(\beta(\mathfrak{a}))$. Proposition 4.9 implies, substituting $t = \theta(S)$,

$$\delta_{k,n}(\beta(\mathfrak{a})) = \left(\frac{\Omega_p}{\lambda'_{\hat{E}}(t)} \frac{d}{dt}\right)^k \log\, g_{e(\mathfrak{a})}(t)|_{t=\theta(\varsigma_n-1)}$$

(26) $$= \Omega_p^k \cdot \left(\frac{d}{dz}\right)^k \log\, \Theta(\Omega - z; L, \mathfrak{a})|_{z=v_n}$$

$$= -12 \cdot \Omega_p^k \cdot E_k(\Omega - v_n; L, \mathfrak{a}) \qquad (cf. 3.1(7))$$

where $v_n$ is the $\mathfrak{p}^n$ division point of $L$ for which $t(\xi(v_n, L)) = \theta(\varsigma_n - 1)$.

**Lemma.** *Let $k \geq 1$, $n \geq 0$, and choose a prime $\mathfrak{q}$, $(\mathfrak{q}, \mathfrak{f}\mathfrak{p}) = 1$, such that $\mathrm{N}\mathfrak{q} \equiv 1 \bmod p^n$, $(\mathfrak{q}, F/K) = (\mathfrak{p}^n, F/K)$. Then*

(27) $$\Omega_p^{-k} \cdot \delta_{k,n}(\beta(\mathfrak{a})) = \Omega^{-k} \cdot (-12)(k-1)! \cdot \varphi^k(\mathfrak{p}^n) \cdot$$
$$\cdot \left\{\mathrm{N}\mathfrak{a} L\left(\overline{\varphi}^k, k; \left(\frac{F_n/K}{\mathfrak{q}}\right)\right) - \varphi^k(\mathfrak{a}) L\left(\overline{\varphi}^k, k; \left(\frac{F_n/K}{\mathfrak{q}\mathfrak{a}}\right)\right)\right\}.$$

*Here $L\left(\chi, s; \left(\frac{M/K}{\mathfrak{c}}\right)\right) = \sum \chi(\mathfrak{a})\mathrm{N}\mathfrak{a}^{-s}$, the sum extending over all integral $\mathfrak{a}$ such that $(\mathfrak{a}, \mathfrak{f}_{M/K}) = 1$, $(\mathfrak{a}, M/K) = (\mathfrak{c}, M/K)$. Both sides of (27) lie in $F_n$.*

PROOF: <u>Step 1</u>. $E_k(\Omega - v_n; L, \mathfrak{a}) = \Lambda(\mathfrak{p}^n)^k \cdot E_k(\Omega; \mathfrak{p}^n L, \mathfrak{a})^{\sigma_\mathfrak{q}}$. Indeed,

$$\Lambda(\mathfrak{p}^n)^k \cdot E_k(\Omega; \mathfrak{p}^n L, \mathfrak{a})^{\sigma_\mathfrak{q}} = (\Lambda(\mathfrak{p}^{-n})^{-k} E_k(\Omega; \mathfrak{p}^n L, \mathfrak{a}))^{\sigma_\mathfrak{q}}$$
$$= E_k(\Lambda(\mathfrak{p}^{-n})\Omega; \Lambda(\mathfrak{p}^{-n})\mathfrak{p}^n L, \mathfrak{a})^{\sigma_\mathfrak{q}} \qquad \text{(proposition 3.3)}$$
$$= E_k(\Lambda(\mathfrak{q}\mathfrak{p}^{-n})\Omega; L, \mathfrak{a}), \qquad \text{since } \Lambda(\mathfrak{q}\mathfrak{p}^{-n})\mathfrak{p}^n\mathfrak{q}^{-1} = \mathcal{O}_K.$$

However, from (12) and (11) we obtain $v_n \equiv \Lambda(\mathfrak{q}\mathfrak{p}^{-n})u_n \ mod\ L$, so from the definition of $u_n$ and $w_n$ (4.4), $\Lambda(\mathfrak{q}\mathfrak{p}^{-n})\Omega \equiv \Lambda(\mathfrak{q}\mathfrak{p}^{-n})w_n - v_n$ modulo $L$. Finally, since $w_n \equiv \Omega \ mod\ L$ and $\Lambda(\mathfrak{q}\mathfrak{p}^{-n}) = \varphi(\mathfrak{q}\mathfrak{p}^{-n}) \equiv 1 \ mod\ \mathfrak{f}$, we conclude that $\Lambda(\mathfrak{q}\mathfrak{p}^{-n})\Omega \equiv \Omega - v_n \ mod\ L$, as desired.

Step 2. From the first step and (26) we see that it remains to prove

$$\Lambda(\mathfrak{p}^n)^k E_k(\Omega;\mathfrak{p}^n L,\mathfrak{a})^{\sigma_\mathfrak{q}} = \Omega^{-k}(k-1)!\varphi^k(\mathfrak{p}^n)\cdot$$
$$\cdot\left\{\mathbf{N}\mathfrak{a}L\left(\overline{\varphi}^k,k;\left(\frac{F_n/K}{\mathfrak{q}}\right)\right)-\varphi^k(\mathfrak{a})L\left(\overline{\varphi}^k,k;\left(\frac{F_n/K}{\mathfrak{a}\mathfrak{q}}\right)\right)\right\}.$$

Writing $\Lambda(\mathfrak{p}^n) = \Lambda(\mathfrak{q}) \cdot \varphi(\mathfrak{p}^n\mathfrak{q}^{-1})$, and using the relation

$$E_k(\Omega;\mathfrak{p}^n L,\mathfrak{a}) = \mathbf{N}\mathfrak{a} \cdot E_k(\Omega,\mathfrak{p}^n L) - \Lambda(\mathfrak{a})^k \cdot E_k(\Omega,\mathfrak{p}^n L)^{\sigma_\mathfrak{a}}$$

(derived from (5) and (8) of §3), we see that it is enough to prove

$$\Lambda(\mathfrak{c})^k \cdot E_k(\Omega,\mathfrak{p}^n L)^{\sigma_\mathfrak{c}} = \Omega^{-k}(k-1)!\varphi^k(\mathfrak{c}) \cdot L\left(\overline{\varphi}^k,k;\left(\frac{F_n/K}{\mathfrak{c}}\right)\right), \tag{28}$$

for any integral ideal $\mathfrak{c}$, $(\mathfrak{c},\mathfrak{f}\mathfrak{p}) = 1$. This was done in proposition 3.5.

**4.11** We can now formulate the first main theorem. Let $\mathfrak{f}$ be an ideal of $K$ with $w_\mathfrak{f} = 1$, and $\mathfrak{p}$ a split prime, $(\mathfrak{p},\mathfrak{f}) = 1$. Let $e_n(\mathfrak{a}) = \Theta(1;\mathfrak{f}\mathfrak{p}^n,\mathfrak{a})$, $(\mathfrak{a},\mathfrak{f}\mathfrak{p}) = 1$, $n \geq 1$, be the elliptic units (23) in $F_n = K(\mathfrak{f}\mathfrak{p}^n)$. Define $e(\mathfrak{a})$ and $\beta(\mathfrak{a})$ as in 4.9 and recall that $\mathcal{G} = Gal(F_\infty/K)$. Let $\mu_\mathfrak{a} = \mu^0_{\beta(\mathfrak{a})}$ be the $p$-adic integral measure on $\mathcal{G}$ corresponding to $\beta(\mathfrak{a})$, as in 4.6-4.7.

**Theorem.** *There exist complex and $p-adic$ "periods" $\Omega \in \mathbf{C}^x$ and $\Omega_p \in \mathbf{C}_p^x$, for which the following interpolation formula, both sides of which lie in $\overline{\mathbf{Q}}$, holds. The grossencharacter $\varepsilon$ is assumed to be of type $(k,0)$, $k \geq 1$, and of conductor dividing $\mathfrak{f}\mathfrak{p}^\infty$.*

$$\Omega_p^{-k}\cdot\int_\mathcal{G}\varepsilon(\sigma)d\mu_\mathfrak{a}(\sigma) = \tag{29}$$
$$\Omega^{-k}12(k-1)!\cdot G(\varepsilon)\left(1-\frac{\varepsilon(\mathfrak{p})}{p}\right)\cdot(\varepsilon(\mathfrak{a})-\mathbf{N}\mathfrak{a})\cdot L_\mathfrak{f}(\varepsilon^{-1},0).$$

*Here the complex L function is taken with modulus* $\mathfrak{f}$. *The "like Gauss sum"* $G(\varepsilon)$ *is defined as follows. Let* $F' = K(\mathfrak{f}\bar{\mathfrak{p}}^{\infty})$, *so that* $F'F_n = K(\mathfrak{f}\mathfrak{p}^n\bar{\mathfrak{p}}^{\infty})$. *Write* $\varepsilon = \chi\varphi^k$ *with a grossencharacter* $\varphi$ *of type (1,0) whose conductor divides* $\mathfrak{f}$. *Let* $S = \{\gamma \in Gal(F'F_n/K) \mid \gamma|F' = (\mathfrak{p}^n, F'/K)\}$ *where n is the exact power of* $\mathfrak{p}$ *dividing the conductor of* $\varepsilon$. *Then*

$$(30) \qquad G(\varepsilon) = \frac{\varphi^k(\mathfrak{p}^n)}{p^n} \cdot \sum_{\gamma \in S} \chi(\gamma)(\varsigma_n^{\gamma})^{-1}.$$

REMARKS: (i) $G(\varepsilon)$ is well defined because $\varsigma_n \in F'F_n$, and it depends only on $\varepsilon$ as a whole (but not on $\varphi$ or $\chi$). $G(\varepsilon) = 1$ if $\varepsilon$ is unramified at $\mathfrak{p}$, and it is an ordinary Gauss sum if $k = 0$. It lies in a CM field ([Go-Sch] §4), and $G(\varepsilon)\overline{G(\varepsilon)} = p^{n(k-1)}$.

(ii) Note that $\mathfrak{a}$ intervenes in the right hand side of (29) only through the twisting factor $\varepsilon(\mathfrak{a}) - \mathbf{N}\mathfrak{a}$.

(iii) The proof shows that we may take $\Omega_p$ to be a unit in the completion of $\mathbf{Q}_p^{ur}$.

(iv) Both the *period-pair-class* $<\Omega, \Omega_p> \in (\mathbf{C}^{x} \times \mathbf{C}_p^{x})/\overline{\mathbf{Q}}^{x}$ *and* $\mu_{\mathfrak{a}}$ are uniquely determined by (29). Indeed, suppose $\tilde{\Omega}$, $\tilde{\Omega}_p$ and $\tilde{\mu}$ also satisfy it. Let $d\lambda_1(\sigma) = \varphi(\sigma)d\mu_{\mathfrak{a}}(\sigma)$ and $d\lambda_2(\sigma) = \varphi(\sigma)d\tilde{\mu}(\sigma)$. Then there exists a constant $c \in \overline{\mathbf{Q}}^{x}$ such that for any $\chi$ of finite order $\int \chi d\lambda_1 = c\int \chi d\lambda_2$. Hence $\lambda_1 = c\lambda_2$, and $\mu_{\mathfrak{a}} = c\tilde{\mu}$. But $c\Omega\tilde{\Omega}^{-1} = \Omega_p\tilde{\Omega}_p^{-1}$, so if (29) is supposed to hold for all $\varphi^k\chi$, $k \geq 1$, then $c = 1$.

This is perhaps the point to remark that the dependence of $\mu_{\mathfrak{a}}$, $\Omega_p$ and $G(\varepsilon)$ on $(\varsigma_n)$ is compatible with (29).

PROOF: Write $\varepsilon = \chi\varphi^k$ as above. Fix $E$ as in 4.1-4.2, $\Omega$ and $\Omega_p$ as in 4.2 (6) and 4.3-4.4, and compute (29) from formulas (16), (21), and (27). Use (2.4(ii)) $\sigma_{\mathfrak{c}}(\beta(\mathfrak{a})) = \beta(\mathfrak{a}\mathfrak{c})\beta(\mathfrak{c})^{-\mathbf{N}\mathfrak{a}}$. We arrive at (29), with $\varphi^k(\mathfrak{p}^n)\chi(\mathfrak{q})\tau(\chi)$ for $G(\varepsilon)$ (see (19)). Now let $F_n' = F(E[\bar{\mathfrak{p}}^n])$. Then $F_nF_n' = K(\mathfrak{f}p^n)$ so $\varsigma_n \in F_nF_n'$ (this is also a consequence of Weil's pairing). We may choose $\mathfrak{q}$ so that $(\mathfrak{q}, F_n'/K) = (\mathfrak{p}^n, F_n'/K)$, and $\mathbf{N}\mathfrak{q} = 1 \; mod \; p^n$. Then if $\gamma \in Gal(F_nF_n'/K)$ and $\gamma|F_n' = (\mathfrak{q}, F_n'/K)$,

$$\varsigma_n^{\gamma} = \varsigma_n^{\gamma\sigma_{\mathfrak{q}}^{-1}} = \varsigma_n^{\kappa(\gamma\sigma_{\mathfrak{q}}^{-1})}$$

because $\gamma\sigma_{\mathfrak{q}}^{-1}$ fixes $E[\overline{\mathfrak{p}}^n]$ pointwise, so its action on $E[\mathfrak{p}^n]$ and $\mu_{p^n}$ is given by the same character. With this in mind, (30) follows.

**4.12 Theorem.** *(i) Let $\mathfrak{f}$ be any non-trivial integral ideal of $K$, and $\mathfrak{p}$ a split prime $(\mathfrak{p}, \mathfrak{f}) = 1$. Let $G(\varepsilon)$ be defined as above (30). Then there exist periods $\Omega \in \mathbf{C}^x$ and $\Omega_p \in \mathbf{C}_p^x$, and a unique $p$-adic integral measure $\mu(\mathfrak{f})$ on $\mathcal{G}(\mathfrak{f}) = Gal(K(\mathfrak{f}\mathfrak{p}^\infty)/K)$, such that for any grossencharacter $\varepsilon$ of conductor dividing $\mathfrak{f}\mathfrak{p}^\infty$ and type $(k,0)$, $k \geq 1$,*

$$(31) \qquad \Omega_p^{-k} \int_{\mathcal{G}(\mathfrak{f})} \varepsilon(\sigma) d\mu(\mathfrak{f};\sigma) = \Omega^{-k} \cdot G(\varepsilon)\left(1 - \frac{\varepsilon(\mathfrak{p})}{p}\right) \cdot L_{\infty,\mathfrak{f}}(\varepsilon^{-1},0).$$

*(ii) If $\mathfrak{f}|\mathfrak{g}$ and $\overline{\mu}(\mathfrak{g})$ is the measure induced from $\mu(\mathfrak{g})$ on $\mathcal{G}(\mathfrak{f})$, then*

$$(32) \qquad \overline{\mu}(\mathfrak{g}) = \prod(1 - \sigma_{\mathfrak{l}}^{-1}) \cdot \mu(\mathfrak{f})$$

*where the product is over all $\mathfrak{l}$ dividing $\mathfrak{g}$ but not $\mathfrak{f}$.*

*(iii) If $\mathfrak{f} = (1)$ the same conclusion holds, except that now $\mu(1)$ is just a pseudo-measure, but for any $\sigma \in \mathcal{G}(1) = Gal(K(\mathfrak{p}^\infty)/K)$, $(1 - \sigma)\mu(1)$ is a $p-$adic integral measure.*

REMARKS: (i) In view of (ii) the theorem remains valid if $\mathfrak{f}$ is replaced by $\mathfrak{f}\mathfrak{g}^\infty$ with $\mathfrak{f}$, $\mathfrak{g}$ prime to $\mathfrak{p}$. An important case occurs when $\overline{\mathfrak{p}}|\mathfrak{g}$, because then grossencharacters of *any* infinity type $(k,j)$ may be integrated, and when $(k,j)$ is critical, a formula similar to (31) holds. See 4.14.

(ii) The grossencharacters for which (31) applies are exactly those that can be interpreted as $p$-adic characters of $\mathcal{G}(\mathfrak{f})$, and are also critical.

(iii) $L_{\infty,\mathfrak{f}}$ refers to the complex $L$ function *with* the Euler factor at $\infty$ (1.1 (2)) but *without* the Euler factors at the primes dividing $\mathfrak{f}$.

(iv) Just as in 4.11, $\mu(\mathfrak{f})$ and $< \Omega, \Omega_p > \in (\mathbf{C}^x \times \mathbf{C}_p^x)/\overline{\mathbf{Q}}^x$ are uniquely determined by (31). In view of (i), the period pair class $< \Omega, \Omega_p >$ is even independent of $\mathfrak{f}$, although the individual periods depend on $\mathfrak{f}$ and on the Weierstrass model of $E$ chosen in the construction.

(v) The (possible) pole at the trivial character—this is the meaning of part (iii)—is a common feature of $p$-adic $L$ functions. Compare [Iw2], p. 29 for the Kubota-Leopoldt $L$ function. We shall prove later (5.3) that $\mu(1)$ indeed has a pole at the trivial character.

PROOF: We note first that (ii) follows from (31), and implies (iii) as well. Simply choose any prime $\mathfrak{l}$, and let $\mu(1) = \overline{\mu}(\mathfrak{l})/(1 - \sigma_{\mathfrak{l}}^{-1})$. We may also assume that $w_{\mathfrak{f}} = 1$, because if $\mathfrak{f}$ fails to satisfy it, some power of it will, and we may let $\mu(\mathfrak{f})$ be the measure induced from $\mu(\mathfrak{f}^m)$.

So fix $\mathfrak{f}$ as in 4.11, and let $\delta_{\mathfrak{a}} = \sigma_{\mathfrak{a}} - \mathbf{N}\mathfrak{a}$ be the "twisting measure" associated with $\mathfrak{a}$. By (29),

$$(33) \qquad \mu_{\mathfrak{a}} \cdot \delta_{\mathfrak{b}} = \mu_{\mathfrak{b}} \cdot \delta_{\mathfrak{a}}, \qquad (\mathfrak{a}\mathfrak{b}, \mathfrak{f}\mathfrak{p}) = 1,$$

because the integrals of the two against any admissible $\varepsilon$ are equal, and there are enough admissible $\varepsilon$ to separate measures apart (grossencharacters $\varphi\chi$ with a fixed $\varphi$, and $\chi$ ranging over all the characters of finite order, already accomplish it). We shall prove, roughly speaking, that the greatest common divisor of all the $\delta_{\mathfrak{a}}$ is 1, hence the pseudo-measures $\mu_{\mathfrak{a}}/\delta_{\mathfrak{a}}$, which are all equal, are actually measures.

Let $K_\infty$ be the maximal $\mathbf{Z}_p$ extensions of $K$ inside $K(\mathfrak{f}\mathfrak{p}^\infty)$, and $G' = Gal(K(\mathfrak{f}\mathfrak{p}^\infty)/K_\infty)$, a finite group, $|G'| = m$. Let $\Gamma' = Gal(K_\infty/K)$, and fix an isomorphism $\mathcal{G} \cong \Gamma' \times G'$. Let $D$ be the ring generated over the integers of the completion of the maximal unramified extension of $K_{\mathfrak{p}}$, by the $m^{th}$ roots of unity. Then $\mu_{\mathfrak{a}} \in D[[\mathcal{G}]] \cong D[[\Gamma']][G']$, and $\mathbf{Q} \otimes D[[\mathcal{G}]] \cong \mathbf{Q} \otimes D[[\Gamma']]^m$. The last isomorphism is through $\lambda \mapsto (\ldots, \theta(\lambda), \ldots)$ where $\theta$ runs over the characters of $G'$, extended to homomorphisms $D[[\mathcal{G}]] \to D[[\Gamma']]$. It is also well known that $D[[\Gamma']] \cong D[[X]]$, and this is a unique factorization domain, since $D$ is a discrete valuation ring.

Now for any $\theta$, $\theta(\delta_{\mathfrak{a}}) = \theta(\sigma_{\mathfrak{a}}|G') \cdot (\sigma_{\mathfrak{a}}|\Gamma') - \mathbf{N}\mathfrak{a}$ is non-zero since $\sigma_{\mathfrak{a}}|\Gamma' \neq 1$. Thus $\delta_{\mathfrak{a}}$ is a non-zero-divisor in $D[[\mathcal{G}]]$. Furthermore, for a fixed $\theta$, the greatest common divisor of $\theta(\delta_{\mathfrak{a}})$ is 1. To see this, observe first that $\pi$, the uniformizer of $D$, does not divide $\theta(\delta_{\mathfrak{a}})$. Secondly, if $\zeta_n \in K(\mathfrak{f}\mathfrak{p}^\infty)$, but $\zeta_{n+1} \notin K(\mathfrak{f}\mathfrak{p}^\infty)$, then for any $\tau \times \sigma \in \Gamma' \times G' = \mathcal{G}$ fixing $\zeta_n$, and any $u \in 1 + p^n\mathbf{Z}_p$, we can find $\mathfrak{a}$'s such that $\theta(\delta_\mathfrak{a})$ converge to $\theta(\sigma)\tau - u$. A common divisor of all the $\theta(\delta_\mathfrak{a})$ must therefore divide (in $D[[X]]$) $\theta(\sigma)(1 + X)^a - u$ for all $u \in 1 + p^n\mathbf{Z}_p$, $a \in p^n\mathbf{Z}_p$, for some $n$. This being impossible, our assertion is proved.

Applying $\theta$ to (33) we conclude that there exists $\mu_\theta \in D[[\Gamma']]$ such that $\theta(\delta_\mathfrak{a}) \cdot \mu_\theta = \theta(\mu_\mathfrak{a})$ for any $\mathfrak{a}$. Let $e_\theta = m^{-1} \cdot \sum_{\sigma \in G'} \theta(\sigma)\sigma^{-1}$ be the idempotent corresponding to $\theta$. Then $\mu = \sum \mu_\theta e_\theta$ is a $p$-adic measure, $\delta_\mathfrak{a} \cdot \mu = \mu_\mathfrak{a}$ for any $\mathfrak{a}$, and $m \cdot \mu$ is *integral.*

Even if $m$ is divisible by $p$, we conclude from this that $\mu$ is integral, as follows. Suppose not, and let $D^\circ$ be the maximal ideal of $D$. Write $\equiv$ to denote congruence modulo $D^\circ[[\mathcal{G}]]$. We can find a scalar multiple $\nu$ of $\mu$, $\nu \in D[[\mathcal{G}]]$, $\nu \not\equiv 0$, but $\delta_\mathfrak{a} \cdot \nu \equiv 0$ for all $\mathfrak{a}$. Let $\nu = \sum_{\sigma \in G'} \nu_\sigma \sigma$, $\nu_\sigma \in D[[\Gamma']]$, and assume without loss of generality that $\nu_1 \not\equiv 0$. Then

$$\delta_\mathfrak{a} \cdot \nu = \sum_{\sigma \in G'} (\nu_{\rho\sigma}\sigma_\mathfrak{a}|\Gamma' - \mathbf{N}\mathfrak{a}\nu_\sigma)\sigma, \qquad \rho = (\sigma_\mathfrak{a}|G')^{-1}.$$

Thus $\nu_{\rho\sigma} \equiv \nu_\sigma \mathbf{N}\mathfrak{a}(\sigma_\mathfrak{a}|\Gamma')^{-1}$ for all $\sigma \in G'$. If $\rho^d = 1$, $\nu_1(1 - (\mathbf{N}\mathfrak{a}(\sigma_\mathfrak{a}|\Gamma')^{-1})^d) \equiv 0$. Since $\nu_1 \not\equiv 0$, $\mathbf{N}\mathfrak{a}^d \equiv (\sigma_\mathfrak{a}|\Gamma')^d$ in $D[[\Gamma']]$, which is a contradiction.

We conclude that $\mu_\mathfrak{a}/\delta_\mathfrak{a} = \mu$ is an integral measure independent of $\mathfrak{a}$. We claim that $\mu(\mathfrak{f}) = \mu/12$ is also integral. When $(p, 6) = 1$ there is nothing to prove. Otherwise we appeal to the results of Robert and Gillard (prop. 2.7). It is easy to deduce from them that if $(\mathfrak{a}, 6\mathfrak{f}\mathfrak{p}) = 1$, then $\beta(\mathfrak{a})$ is a $12^{th}$ power in $\mathcal{U}$, hence $\mu_\mathfrak{a}$ is divisible by 12. Replace $\mu_\mathfrak{a}$ by $\mu_\mathfrak{a}/12$ and repeat the arguments above. The claim follows. Comparing (29) with (31) concludes the proof.

**4.13** As mentioned above, of special interest is the case when $\mathfrak{f}$ is replaced by $\mathfrak{f}\overline{\mathfrak{p}}^\infty$. Then $\mathcal{G} = Gal(K(\mathfrak{f}p^\infty)/K)$ contains the unique $\mathbf{Z}_p^2$-extension of $K$, and *any* grossencharacter of type $A_0$ and conductor dividing $\mathfrak{f}p^\infty$ may be regarded as a $p$-adic character of $\mathcal{G}$. Such grossencharacters can now be integrated against the measure produced in 4.12 on $\mathcal{G}$, and theorem 4.14 will give the interpolation property extending (31).

One point is nevertheless different. A glance at (36) below reveals that its validity is not insensitive anymore to the choice of $(\zeta_n)$. If $\gamma \in Gal(K(\mathfrak{f}p^\infty)/K(\mathfrak{f}\overline{\mathfrak{p}}^\infty))$ and $\tilde{\zeta}_n = \zeta_n^\gamma$, then the new $\mu$, $\Omega_p$ and $G(\varepsilon)$, for $\varepsilon = \chi\varphi^k\overline{\varphi}^j$ with $(\mathfrak{f}_\varphi, \mathfrak{p}) = 1$, are given by

$$(34) \qquad \tilde{\mu}(U) = \mu(\gamma U), \qquad \tilde{\Omega}_p = \mathbf{N}\gamma^{-1} \cdot \Omega_p, \qquad \tilde{G}(\varepsilon) = \chi(\gamma)^{-1} \cdot G(\varepsilon).$$

Note that $\overline{\varphi}(\gamma) = 1$, since $\gamma$ fixes $E[\overline{\mathfrak{p}}^\infty]$ pointwise, so $\varphi(\gamma) = \mathbf{N}\gamma$. Formula (36) is insensitive to such a change only if $j = 0$. We shall therefore begin by prescribing the $(\zeta_n)$ for which the theorem holds. However, a few words of explanation are probably needed.

There are actually *two* $p$-adic periods associated to $\varphi$, $\Omega_p'$ and $\Omega_p''$, which intrinsically lie in $\mathbf{C}_p(1)$ and $\mathbf{C}_p$, and satisfy $\Omega_p' \cdot \Omega_p'' = 1 \otimes 2\pi i$. Both are units in the completion of the maximal unramified extension of $\mathbf{Q}_p$ (with the right Tate twist), and Galois acts on them like

$$(10') \qquad \sigma_\mathfrak{c}(\Omega_p') = \Lambda(\mathfrak{c})\Omega_p',$$

$$(10'') \qquad \sigma_\mathfrak{c}(\Omega_p'') = \mathbf{N}\mathfrak{c}\Lambda(\mathfrak{c})^{-1}\Omega_p''.$$

Here, by choosing an appropriate orientation $(\zeta_n)$, we shall identify $\mathbf{C}_p(1)$ with $\mathbf{C}_p$ so that $1 \otimes 2\pi i$ corresponds to 1, $\Omega_p' = \Omega_p \otimes 2\pi i$, and $\Omega_p'' = \Omega_p^{-1}$. The transcendental part of $\int \varphi^k\overline{\varphi}^j d\mu$, intrinsically given by $\Omega_p'^k\Omega_p''^j$, becomes $\Omega_p^{k-j}$, which accounts for (36) below. See [dS3] for generalizations to CM fields, and for the relation with Hodge theory of abelian varieties.

CONVENTION. *From now on, let $\zeta_n$ be the primitive $p^n$ root of unity given by the Weil pairing (see [La] ch. 18, [Mum] §20)*

$$(35) \qquad \zeta_n = e_{p^n}(w_n, u_n)$$

*with respect to the lattice $\mathfrak{p}^n\mathcal{O}_K\Omega$, where $w_n$ and $u_n$ are the unique $\overline{\mathfrak{p}}^n$– and $\mathfrak{p}^n$–division points of that lattice for which $w_n - u_n \equiv \Omega$ mod $p^n\mathcal{O}_K\Omega$.*

The following facts are easy to establish:

(i) $\varsigma_n^p = \varsigma_{n-1}$.

(ii) If $\mathfrak{f} = \bar{\mathfrak{p}}^m$ and $0 \le n \le m$, and if $w_n$ and $u_n$ are defined as in 4.4, then $\varsigma_n = e_{p^m}(w_n, u_n)$ with respect to the lattice $\mathfrak{p}^n\mathfrak{f}\Omega$. For $n = m$ this agrees, of course, with the definition.

REMARK: $\varsigma_n$ can be given in a closed form. See lemma 6.2.

**4.14 Theorem.** *Let $\mathfrak{g}$ be an integral ideal of $K$, and $p$ a split rational prime, $(p, \mathfrak{g}) = 1$. Let $\mu$ be the measure $\mu(\mathfrak{g}\bar{\mathfrak{p}}^\infty)$ on $\mathcal{G} = Gal(K(\mathfrak{g}p^\infty)/K)$ (see remark (i) to theorem 4.12), and fix $< \Omega, \Omega_p > \in (\mathbf{C}^x \times \mathbf{C}_p^x)/\overline{\mathbf{Q}}^x$ as above. Then the following formula, both sides of which lie in $\overline{\mathbf{Q}}$, holds for any grossencharacter $\varepsilon$ of conductor dividing $\mathfrak{g}p^\infty$, and of type $(k, j)$, $0 \le -j < k$:*

$$(36)\quad \Omega_p^{j-k} \int_{\mathcal{G}} \varepsilon(\sigma) d\mu(\sigma) = \Omega^{j-k} \left(\frac{\sqrt{d_K}}{2\pi}\right)^j \cdot G(\varepsilon) \left(1 - \frac{\varepsilon(\mathfrak{p})}{p}\right) \cdot L_{\infty,\mathfrak{g}\bar{\mathfrak{p}}}(\varepsilon^{-1}, 0).$$

*Here to define $G(\varepsilon)$, write $\varepsilon = \varphi^k \bar{\varphi}^j \chi$ with a grossencharacter $\varphi$ of conductor prime to $\mathfrak{p}$ and type (1,0), and $\chi$ a character of finite order, and let (see (30))*

$$(37)\qquad G(\varepsilon) = \frac{\varphi^k \bar{\varphi}^j(\mathfrak{p}^n)}{p^n} \cdot \sum_{\gamma \in S} \chi(\gamma) \cdot (\varsigma_n^\gamma)^{-1}.$$

PROOF: This is similar to 4.11, except that the computations are more complicated. We outline the main steps, and leave out some routine verifications to the reader.

Step 1. Fix an auxiliary ideal $\mathfrak{a}$, $(\mathfrak{a}, \mathfrak{g}p) = 1$. With the notation of 4.11, and with $\mathfrak{f} = \mathfrak{g}\bar{\mathfrak{p}}^m$, $m \ge 1$, $\mu_\mathfrak{a} = 12(\sigma_\mathfrak{a} - \mathbf{N}\mathfrak{a})\mu(\mathfrak{f})$. Since the measures $\mu(\mathfrak{f})$, for various $m$, are compatible (4.12(ii)), so are $\mu_\mathfrak{a}$, and their inverse limit is a measure $\mu_\mathfrak{a}$ on $\mathcal{G}$. This $\mu_\mathfrak{a}$ is associated to the *double* inverse system of units $\Theta(1, \mathfrak{g}\bar{\mathfrak{p}}^m\mathfrak{p}^n; \mathfrak{a}) = e_{n,m}(\mathfrak{a})\quad (n, m \ge 1)$.

Now fix some large $m \ge 1$, let $\mathfrak{f} = \mathfrak{g}\bar{\mathfrak{p}}^m$, and assign $F$, $E$, $L$, $\Omega$, $\Omega_p$, $\varphi$, $\chi$ and $\Lambda$ their previous meaning, e.g. $L = \mathfrak{f}\Omega$ etc. Let $F_n = K(\mathfrak{f}\mathfrak{p}^n)$ and $G_n = Gal(F_\infty/F_n)$. Let $\mathfrak{c}$ run over integral ideals of $K$ whose Artin symbols represent $Gal(F_n/K)$. Write $\equiv$ to indicate congruences in $\mathbf{C}_p$ (not necessarily between integers). Now, if $n \geq 0$ is the power of $\mathfrak{p}$ in $\mathfrak{f}_\chi$,

$$\Omega_p^{j-k} \cdot \int_{\mathcal{G}} \chi\varphi^k\overline{\varphi}^j(\sigma)\cdot d\mu_{\mathfrak{a}}(\sigma)$$

$$(38) \equiv \Omega_p^{j-k} \cdot \sum_{\mathfrak{c}} \chi\varphi^k\overline{\varphi}^j(\mathfrak{c}^{-1}) \cdot \int_{G_n} \varphi^k(\sigma)\cdot d\mu^0_{\sigma_{\mathfrak{c}}(\beta(\mathfrak{a}))}(\sigma) \quad mod\ p^m$$

$$\equiv \Omega_p^{j-k} \cdot \tau(\chi) \cdot \left(1 - \frac{\chi\varphi^k\overline{\varphi}^j(\mathfrak{p})}{p}\right) \cdot \sum_{\mathfrak{c}} \chi\varphi^k\overline{\varphi}^j(\mathfrak{c}^{-1}) \cdot \delta_{k,n}(\sigma_{\mathfrak{c}}(\beta(\mathfrak{a}))^0 \quad mod\ p^m$$

as in lemmas 4.7 and 4.8. If we write $\sigma_{\mathfrak{c}}(\beta(\mathfrak{a})) = \beta(\mathfrak{a}\mathfrak{c}) \cdot \beta(\mathfrak{c})^{-\mathbf{N}\mathfrak{a}}$, we deduce at once from (26) that (38) is congruent modulo $p^m$ to

$$12 \cdot \Omega_p^j \cdot \tau(\chi) \cdot \left(1 - \frac{\chi\varphi^k\overline{\varphi}^j(\mathfrak{p})}{p}\right) \cdot \sum_{\mathfrak{c}} \chi\varphi^k\overline{\varphi}^j(\mathfrak{c}^{-1}) \cdot$$
$$\cdot \{\mathbf{N}\mathfrak{a} \cdot E_k(\Omega - v_n; L, \mathfrak{c}) - E_k(\Omega - v_n; L, \mathfrak{a}\mathfrak{c})\}.$$

Let $\mathfrak{q}$ be an ideal as in lemma 4.10. As in the first step of that lemma, we conclude that (38) is congruent modulo $p^m$ to

$$(39) \qquad \begin{aligned} &12 \cdot \Omega_p^j \cdot \tau(\chi) \cdot \left(1 - \frac{\chi\varphi^k\overline{\varphi}^j(\mathfrak{p})}{p}\right) \cdot \Lambda(\mathfrak{p}^n\mathfrak{q}^{-1})^k \cdot \sum_{\mathfrak{c}} \chi\varphi^k\overline{\varphi}^j(\mathfrak{c}^{-1}) \cdot \\ &\cdot \{\Lambda(\mathfrak{a}\mathfrak{c}\mathfrak{q})^k E_k(\Omega, \mathfrak{p}^n L)^{\sigma_{\mathfrak{a}\mathfrak{c}\mathfrak{q}}} - \mathbf{N}\mathfrak{a} \cdot \Lambda(\mathfrak{c}\mathfrak{q})^k E_k(\Omega, \mathfrak{p}^n L)^{\sigma_{\mathfrak{c}\mathfrak{q}}}\}. \end{aligned}$$

Step 2. Recall that $\mu_{\mathfrak{a}} = 12(\sigma_{\mathfrak{a}} - \mathbf{N}\mathfrak{a})\mu$, and $\varepsilon = \chi\varphi^k\overline{\varphi}^j$. We shall prove that for $(\mathfrak{c}, \mathfrak{f}\mathfrak{p}) = 1$,

$$(40) \qquad \begin{aligned} &\Omega_p^j \cdot \Lambda(\mathfrak{p}^n\mathfrak{q}^{-1}\mathfrak{c})^k \cdot E_k(\Omega, \mathfrak{p}^n L)^{\sigma_{\mathfrak{c}}} \equiv \\ &(k-1)! \left(\frac{\sqrt{d_K}}{2\pi}\right)^j \Omega^{j-k} \cdot \varphi^k(\mathfrak{p}^n\mathfrak{q}^{-1}) \cdot \varphi^k\overline{\varphi}^j(\mathfrak{c}) \cdot \\ &\cdot L\left(\overline{\varphi}^{k-j}, k; \left(\frac{F_n/K}{\mathfrak{c}}\right)\right) mod\ 2^j p^{m-m_0} \end{aligned}$$

for some $m_0$ *independent of* $m$. Simple algebra shows that (39) and (40) together imply a congruence modulo $p^{m-m_0}$ between the two sides of (36). Our $m_0$ will depend on $\mathfrak{g}$ and $\varepsilon$, but not on $m$. Since $m$ may be arbitrarily large, the two sides of (36) are in fact equal.

The key to (40) is the congruence

$$\Omega_p^{-1} \equiv -\mathbf{N}(\mathfrak{f}\mathfrak{p}^n)E_1(\Omega, \mathfrak{p}^n\mathfrak{f}\Omega) \qquad mod\ p^{m-m_0} \tag{41}$$

($m_0$ independent of $m$), to be treated in 4.15. Assume (41) is proven, and consider the fundamental relation 3.5(13):

$$\begin{aligned}(k-1)!\left(\frac{\sqrt{d_K}}{2\pi}\right)^j \Omega^{j-k} \cdot \varphi(\mathfrak{c})^{k-j} \cdot L\left(\overline{\varphi}^{k-j}, k; \left(\frac{F_n/K}{\mathfrak{c}}\right)\right) &= \\ \mathbf{N}(\mathfrak{f}\mathfrak{p}^n)^{-j}\Lambda(\mathfrak{c})^{k-j} \cdot E_{j,k}(\Omega, \mathfrak{p}^n L)^{\sigma_\mathfrak{c}},&\end{aligned} \tag{42}$$

of which (28) is a special case. We would like to replace $E_{j,k}$ by an expression involving $E_k$ only, possibly weakening the equality in (42) and replacing it by a congruence. This can be done, according to 3.4 (12), but with $E_{j,k}(\Lambda(\mathfrak{p}^{-n})\Omega, \Lambda(\mathfrak{p}^{-n})\mathfrak{p}^n L)$, because the lattice $\Lambda(\mathfrak{p}^{-n})\mathfrak{p}^n L$ (unlike $\mathfrak{p}^n L$) is the period lattice of a Weierstrass model with good reduction. Indeed, $\Lambda(\mathfrak{p}^{-n})\mathfrak{p}^n L$ is the period lattice of $E^{\phi^{-n}}$, where $E$ refers to the specific model fixed in the beginning, which had good reduction at every place of $F = K(\mathfrak{f})$ above $\mathfrak{p}$. Therefore express the right hand side of (42) as

$$\mathbf{N}(\mathfrak{f}\mathfrak{p}^n)^{-j}\Lambda(\mathfrak{c}\mathfrak{p}^{-n})^{k-j} \cdot E_{j,k}(\Lambda(\mathfrak{p}^{-n})\Omega, \Lambda(\mathfrak{p}^{-n})\mathfrak{p}^n L)^{\sigma_\mathfrak{c}},$$

and use 3.4 (12) to conclude that (42) is congruent modulo $2^j p^{m-nk}$ to

$$\begin{aligned}&\Lambda(\mathfrak{c})^{k-j}(-\mathbf{N}(\mathfrak{f}\mathfrak{p}^n)E_1(\Omega, \mathfrak{p}^n L)^{\sigma_\mathfrak{c}})^{-j}E_k(\Omega, \mathfrak{p}^n L)^{\sigma_\mathfrak{c}} \\ &\equiv \sigma_\mathfrak{c}\Omega_p^j \cdot \Lambda(\mathfrak{c})^{k-j} \cdot E_k(\Omega, \mathfrak{p}^n L)^{\sigma_\mathfrak{c}} \qquad mod\ 2^j p^{m-m_0-nk} \qquad \text{by (41)} \\ &\equiv \Omega_p^j \mathbf{N}\mathfrak{c}^{-j}\Lambda(\mathfrak{c})^k \cdot E_k(\Omega, \mathfrak{p}^n L)^{\sigma_\mathfrak{c}} \qquad mod\ 2^j p^{m-m_0-nk} \qquad \text{by (10)}.\end{aligned}$$

To deduce (40), multiply the left hand side of (42) and the last line by $\mathbf{N}\mathfrak{c}^j\Lambda(\mathfrak{p}^n\mathfrak{q}^{-1})^k$, and recall that $\Lambda(\mathfrak{p}^n\mathfrak{q}^{-1}) = \varphi(\mathfrak{p}^n\mathfrak{q}^{-1})$ because $\mathfrak{p}^n\mathfrak{q}^{-1}$ is a principal ideal congruent to 1 modulo $\mathfrak{f}$.

REMARK: It is plausible that $m_0 = 0$ satisfies (41) and (40). See the remark following lemma 3.4. On the other hand, the extra powers of 2 (when $p = 2$) seem to be necessary. One way or another, the final result is unaffected by this, because our concern is with the limit when $m$ approaches $\infty$.

**4.15** STEP 3. PROOF OF THE CONGRUENCE (41): The idea behind the proof is simple. Specifying an isomorphism $\theta : \hat{\mathbf{G}}_m \simeq \hat{E}$ is tantamount to giving an isomorphism between the corresponding $p$-divisible groups, and modulo $p^m$, $\theta'(0)$ is determined by the restriction of that isomorphism to the group schemes $\mu_{p^m} \simeq E[\mathfrak{p}^m]$. However, by the Weil pairing, this is the same as giving a primitive $\overline{\mathfrak{p}}^m$ division point on $E$. We are therefore led to compute a formula for the Weil pairing in terms of the parameter $t$ on $E$.

First, by lemma 3.4(iii), and at the expense of introducing $m_0$, we may assume that $\mathfrak{g} = (1)$ and prove (41) with $\mathfrak{f} = \overline{\mathfrak{p}}^m$. Let $u_n$ and $w_n$ be, as in 4.4, the unique $\mathfrak{p}^n$- and $\mathfrak{f}$- division points of $\mathfrak{p}^n\mathfrak{f}\Omega$ for which $w_n - u_n \equiv \Omega \; mod \; \mathfrak{p}^n\mathfrak{f}\Omega$. Let $P_n = \xi(\Lambda(\mathfrak{p}^{-n})u_n, \Lambda(\mathfrak{p}^{-n})\mathfrak{p}^n\mathfrak{f}\Omega)$, $Q_n = \xi(\Lambda(\mathfrak{p}^{-n})w_n, \Lambda(\mathfrak{p}^{-n})\mathfrak{p}^n\mathfrak{f}\Omega)$ be the corresponding points of $\phi^{-n}E$. We consider only those $n$ lying between 0 and $m$. As mentioned at the end of 4.13, $\varsigma_n = e_{p^m}(Q_n, P_n)$. Put, for brevity, $L_n = \Lambda(\mathfrak{p}^{-n})\mathfrak{p}^n\mathfrak{f}\Omega$, and let us write $\sigma(z)$ for the corresponding $\sigma(z, L_n)$. Put also $\tau_n = \Lambda(\mathfrak{p}^{-n})u_n$, $\nu_n = \Lambda(\mathfrak{p}^{-n})w_n$, and choose auxiliary torsion points $\lambda_n$ of $L_n$, primitive of order $\ell$, $(\ell, \mathfrak{f}p) = 1$, in such a way that $\xi(\lambda_n, L_n) = \phi^{-n}\xi(\lambda_0, L)$, $0 \le n \le m$.

If $f$ and $g$ are $L_n$-elliptic functions with

$$div(f) = p^m((\nu_n + \lambda_n) - (\lambda_n)), \quad div(g) = p^m((\tau_n) - (0)),$$

then the Weil pairing of $Q_n$ and $P_n$ is computed by

$$e_{p^m}(Q_n, P_n) = \frac{f((\tau_n) - (0))}{g((\nu_n + \lambda_n) - (\lambda_n))}. \tag{43}$$

Let $h_n$ be the $L_n$-elliptic function

$$h_n(z) = \frac{\sigma(z - \lambda_n) \cdot \sigma(\lambda_n + \nu_n - p^m z) \cdot \sigma(\lambda_n + p^m\nu_n)}{\sigma(p^m z - \lambda_n) \cdot \sigma(\lambda_n + \nu_n) \cdot \sigma(\lambda_n + p^m\nu_n - z)}. \tag{44}$$

Expressing $f$ and $g$ in terms of sigma functions we find out that

$$e_{p^m}(Q_n, P_n) = h_n(\tau_n). \tag{45}$$

Now let us expand $h_n$ in terms of the parameter $t$ on $\phi^{-n}E$. Since $dz/dt(0) = 1$, one gets

$$h_n = 1 + \{(p^m - 1)\varsigma(\lambda_n) - p^m\varsigma(\lambda_n + \nu_n) + \varsigma(\lambda_n + p^m\nu_n)\}t + \cdots \tag{46}$$

where $\varsigma = \sigma'/\sigma$ is Weierstrass' zeta function. However, since $p^m \nu_n \in L_n$, $\varsigma(\lambda_n + p^m \nu_n) = \varsigma(\lambda_n) + p^m \eta(\nu_n)$, so the coefficient of $t$ becomes

$$\begin{aligned} p^m(\varsigma(\lambda) - \eta(\lambda)) - p^m(\varsigma(\lambda+\nu) - \eta(\lambda+\nu)) &= p^m(E_1(\lambda_n, L_n) - E_1(\lambda_n + \nu_n, L_n)) \\ &= p^m \cdot \phi^{-n}(E_1(\lambda_0, L) - E_1(\lambda_0 + \nu_0, L)) \\ &\equiv -p^m \cdot \phi^{-n} E_1(\nu_0, L) \quad mod\ p^{m-m_0} \\ &\equiv \phi^{-n}(-\mathrm{N}\mathfrak{f} \cdot E_1(\Omega, \mathfrak{f}\Omega)) \quad mod\ p^{m-m_0}, \end{aligned}$$

where $m_0$ is independent of $m$. The congruence is a consequence of 3.4(ii), and with an appropriate $\lambda_n$ we could reach $m_0 = 0$, but this is irrelevant to our purpose.

It is clear that $h_n$ is rational over $F(E[\ell])$. We claim that its $t$-expansion (46) has $\mathfrak{p}$-integral coefficients. To prove the claim, let

$$(47) \qquad k_n(z) = \prod_{\gamma \in (p^{-m}L_n/L_n)/\pm 1}{}' \left( \frac{\wp\left(\frac{\lambda+\nu}{p^m} - z\right) - \wp(\gamma)}{\wp\left(\frac{\lambda}{p^m} - z\right) - \wp(\gamma)} \right) \cdot \prod_{0 \le i < p^{2m}} \left( \frac{\wp\left(\frac{2\lambda + (i+1)\nu}{2p^m} - z\right) - \wp\left(\frac{(i-1)\nu}{2p^m}\right)}{\wp\left(\frac{2\lambda + (i+1)\nu}{2p^m} - z\right) - \wp\left(\frac{(i+1)\nu}{2p^m}\right)} \right).$$

Here $\wp(z) = \wp(z, L_n)$, $\nu = \nu_n$, $\lambda = \lambda_n$, and the first product is taken over a full set of representatives of non-zero $p^m$-division points of $L_n$ modulo $\pm 1$. This has to be modified when $p = 2$ as follows. If $\gamma_1$, $\gamma_2$, $\gamma_3$ are the three non-zero 2-division points, one should only take $\wp'(\cdot)/2 = \left\{\prod_{i=1}^3 (\wp(\cdot) - \wp(\gamma_i))\right\}^{1/2}$ as their contribution to the first product. Now it is easy to see that $h_n$ and $k_n$ have the same divisors. Furthermore, the same arguments as in the proof of proposition 4.9, or proposition 2.4(iii) ("second proof" there), show that the $t$-expansion of $k_n(z)$ around 0 has $\mathfrak{p}$-integral coefficients. It also shows that $k_n(0)$ is a $\mathfrak{p}$-adic unit. Hence $h_n(z) = k_n(0)^{-1} k_n(z)$ has a $\mathfrak{p}$-integral $t$-expansion, and our claim is verified.

The choice of $\lambda_n$ and $\nu_n$ makes the power series

$$h = \phi^n h_n$$

independent of $n$. It has $\mathfrak{p}$-integral coefficients, and satisfies

$$(48)\qquad \begin{cases} h \equiv 1 + (-\mathbf{N}\mathfrak{f}\cdot E_1(\Omega, \mathfrak{f}\Omega))t + \cdots \ mod\ p^{m-m_0} \\ \varsigma_n = (\phi^{-n}h)(\omega_n) \qquad 0 \le n \le m. \end{cases}$$

The last equality is a consequence of (45) and the specific choice of $(\varsigma_n)$ outlined in 4.13. For $m$ large enough (48) already guarantees that $-\mathbf{N}\mathfrak{f} \cdot E_1(\Omega, \mathfrak{f}\Omega)$ *is a* $\mathfrak{p}$*-adic unit*, and the power series $h - 1$ has an inverse with respect to composition, which we denote by $\tilde{\theta}$. Recall (11) that $\theta$ was the *unique* isomorphism $\hat{\mathbf{G}}_m \simeq E$ for which $\omega_n = (\phi^{-n}\theta)(\varsigma_n - 1)$. Hence $\theta$ and $\tilde{\theta}$ are two power series with coefficients in an unramified extension of $\mathbf{Q}_p$, for which

$$(\phi^{-n}\theta)(\varsigma_n - 1) = (\phi^{-n}\tilde{\theta})(\varsigma_n - 1), \qquad 0 \le n \le m.$$

This implies that $[(1 + X)^{p^n} - 1]/[(1 + X)^{p^{n-1}} - 1]$ divides $\phi^{-n}(\theta - \tilde{\theta})$, hence $\theta - \tilde{\theta}$, for $1 \le n \le m$, so $\theta'(0) \equiv \tilde{\theta}'(0)\ mod\ p^m$. Since $\theta'(0) = \Omega_p$, we finally obtain (41). The proof of theorem 4.14 is complete.

REMARK: Step 3 settles affirmatively a question left open in [Ya2], and also shows how to choose $(\varsigma_n)$ so that the periods denoted by $\Omega_\mathfrak{p}$ in [Ya], and by $\gamma_\mathfrak{p}$ in [Ya2], become equal.

**4.16** $p$-ADIC $L$ FUNCTIONS: For convenience and comparison with results in the literature, we give the translation of our main theorems to power series language.

As usual, the quadratic imaginary field $K$ and the embeddings of $\overline{\mathbf{Q}}$ in $\mathbf{C}$ and $\mathbf{C}_p$ are fixed once and for all, and $p$ is assumed to split in $K$ as $\mathfrak{p}\overline{\mathfrak{p}}$, where $\mathfrak{p}$ is the place induced from $\mathbf{C}_p$. Also fixed throughout the discussion is the generator $(\varsigma_n)$ for the Tate module of $\mu_{p^\infty}$, as in 4.13, and the period-pair-class $< \Omega, \Omega_p > \in (\mathbf{C}^x \times \mathbf{C}_p^x)/\overline{\mathbf{Q}}^x$, as in the preceding sections. We denote by $\varepsilon$ a grossencharacter of $K$ of type $A_0$, or more generally, a $p$-adic character of the Galois group of an abelian extension of $K$.

DEFINITION. *Let $\mathfrak{f}$ be an integral ideal (or a "pseudo-ideal" of the form $\mathfrak{a}\mathfrak{b}^\infty$) relatively prime to $\mathfrak{p}$. The $p$-adic $L$ function of $K$ with modulus $\mathfrak{f}$ is the function whose* *domain is the set of all p-adic continuous characters on* $\mathcal{G}(\mathfrak{f}) = Gal(K(\mathfrak{f}\mathfrak{p}^{\infty})/K)$, *and which assigns to every* $\varepsilon$ ($\varepsilon \neq 1$ *if* $\mathfrak{f} = (1)$) *the value*

$$(49) \qquad L_{p,\mathfrak{f}}(\varepsilon) = \int_{\mathcal{G}(\mathfrak{f})} \varepsilon^{-1}(\sigma) d\mu(\mathfrak{f};\sigma).$$

*Here* $\mu(\mathfrak{f})$ *is the integral measure constructed in theorem 4.12 (see also remark (i) there).*

With this definition, the interpolation formulas (31) and (36) say the following. Let $\varepsilon$ be a grossencharacter of type $(k, j)$ with $0 \leq j < -k$. Define $G(\varepsilon)$ by (37). Let $\mathfrak{g}$ be relatively prime to $p$, and divisible by the non-$p$ part of the conductor of $\varepsilon$. Let $\mathfrak{f} = \mathfrak{g}\overline{\mathfrak{p}}^{\infty}$, and $L_{\infty,\mathfrak{f}}(\varepsilon, s)$ the classical $L$ function *with* the right gamma factor (1.1) but *without* the Euler factors at the primes dividing $\mathfrak{f}$. Then

$$(50) \qquad \Omega_p^{k-j} L_{p,\mathfrak{f}}(\varepsilon) = \Omega^{k-j}\left(\frac{2\pi}{\sqrt{d_K}}\right)^j \cdot G(\varepsilon^{-1})\left(1 - \frac{\varepsilon^{-1}(\mathfrak{p})}{p}\right) \cdot L_{\infty,\mathfrak{f}}(\varepsilon, 0).$$

Furthermore, if $j = 0$ and $\varepsilon$ is unramified at $\overline{\mathfrak{p}}$, the same formula holds with $\mathfrak{g}$ instead of $\mathfrak{f}$.

**4.17** ONE AND TWO-VARIABLE $p$-ADIC $L$ FUNCTIONS: Let $K_{\infty}$ be the unique $\mathbf{Z}_p$ extension of $K$ unramified outside $\mathfrak{p}$. Fix an isomorphism $\kappa_1 : Gal(K_{\infty}/K) \simeq 1 + p\mathbf{Z}_p$ ($1 + 4\mathbf{Z}_2$ if $p = 2$). For example, if $p$ does not divide the class number of $K$, $\mathfrak{p}$ is totally ramified in $K_{\infty}$, and $\kappa_1$ may be taken to be the inverse of the local Artin map, once we identify the Galois group with the inertia group at $\mathfrak{p}$. Let $F$ be an abelian extension of $K$, linearly disjoint from $K_{\infty}$, and $F_{\infty} = FK_{\infty}$ (this notation is somewhat different from the one adopted until now). Fix an isomorphism $Gal(F_{\infty}/K) \simeq Gal(F/K) \times Gal(K_{\infty}/K)$. If $\chi$ is a character of finite order of $Gal(F_{\infty}/K)$, then we may define

$$(51) \qquad L_{p,\mathfrak{f}}(\chi, s) = L_{p,\mathfrak{f}}(\chi\kappa_1^{-s}) \qquad \forall\, s \in \mathbf{Z}_p$$

($\mathfrak{f}$ must be divisible by the non-$\mathfrak{p}$ part of the conductor of $\chi$). Let $\gamma_0$ be a topological generator of $Gal(K_{\infty}/K)$, $\kappa_1(\gamma_0) = u$, and decompose $\chi$ as $\chi = \chi_0\chi_1$, where $\chi_0$

is trivial on $Gal(K_\infty/K)$ (a character "of the first kind", at most tamely ramified at $\mathfrak{p}$), and $\chi_1$ is trivial on $Gal(F/K)$ (in general wildly ramified at $\mathfrak{p}$). Then the fact that $L_{p,\mathfrak{f}}(\chi\kappa_1^{-s})$ is given by an integral like (49) translates to the following statement. There exists a power series $G(\chi_0;T) \in \mathbf{D}[[T]]$ such that

$$L_{p,\mathfrak{f}}(\chi,s) = G(\chi_0;\chi_1(\gamma_0)^{-1}u^s - 1). \tag{52}$$

If $\mathfrak{f} = (1)$ and $\chi_0 = 1$, $G(\chi_0;T) \in T^{-1}\mathbf{D}[[T]]$. Indeed, we may assume $F \subset K(\mathfrak{f})$, so that $F_\infty \subset K(\mathfrak{f}\mathfrak{p}^\infty)$ and $Gal(F_\infty/K)$ is a quotient of $\mathcal{G}(\mathfrak{f}) = Gal(K(\mathfrak{f}\mathfrak{p}^\infty)/K)$. Fix an isomorphism $\mathcal{G}(\mathfrak{f}) \simeq Gal(K_\infty/K) \times Gal(K(\mathfrak{f}\mathfrak{p}^\infty)/K_\infty) = \Gamma' \times H'$, and put

$$G(\chi_0;T) = \sum_{\tau\in H'} \chi_0^{-1}(\tau) \int_{\mathbf{Z}_p} (1 + T)^a d\mu(\mathfrak{f};\tau\gamma_0^a). \tag{53}$$

One often refers to $G(\chi_0;T)$ as the $p$-adic $L$ function of $\chi_0$, and more generally to $G(\chi;T) = G(\chi_0;\chi_1(\gamma_0)^{-1}(1+T) - 1)$ as the $p$-adic $L$ function of $\chi$ (with modulus $\mathfrak{f}$). Note however that as a power series in $T$ it depends on the choice of $u$ (or $\gamma_0$), and only $L_{p,\mathfrak{f}}(\chi,s)$ has an intrinsic meaning.

The two-variable functions are defined by the same procedure. Let $K'_\infty$ be the unique $\mathbf{Z}_p$ extension unramified outside $\overline{\mathfrak{p}}$, and assume for simplicity that $K'_\infty \cap K_\infty = K$. Fix an isomorphism $\kappa_2 : Gal(K'_\infty/K) \simeq 1 + p\mathbf{Z}_p$ ($1 + 4\mathbf{Z}_2$ if $p = 2$). Then we let

$$L_{p,\mathfrak{f}}(\chi;s_1,s_2) = L_{p,\mathfrak{f}}(\chi\kappa_1^{-s_1}\kappa_2^{-s_2}) \qquad \forall\, s_1,s_2 \in \mathbf{Z}_p. \tag{54}$$

We leave it to the reader to define $G(\chi;T_1,T_2)$ as a power series in two variables for any character of finite order $\chi$ of $K$, and to derive the analogues of (52) and (53).

## 5. A $p$-ADIC ANALOGUE OF KRONECKER'S LIMIT FORMULA

**5.1** In the cyclotomic case, Leopoldt has shown that the value of the (Kubota-Leopoldt) $p$-adic $L$ function at 1 is given by a formula which resembles the classical formula for $L(\chi, 1)$. The interest in his result is that the point $s = 1$ is the first integral point *outside* the range of interpolation, yet when the complex logarithm is formally replaced by the $p$-adic one, the same identities hold with $L_p(\chi, 1)$ instead of $L(\chi, 1)$. See [Iw2] theorem 3, p. 61.

When $K$ is quadratic imaginary, there is a similar classical result derived from Kronecker's second limit formula. See [Sie] theorem 9, p. 110. Since the functional equation relates the points $s = 0$ and $s = 1$, the result can be restated as follows ([Ta3] p. 97):

**Theorem.** *Let $\mathfrak{g}$ be a non-trivial integral ideal of $K$, and $\chi$ a character of finite order whose conductor divides $\mathfrak{g}$. Let $\varphi_{\mathfrak{g}}(C)$, $C \in Cl(\mathfrak{g})$, be Robert's invariants associated with classes in the ray class group modulo $\mathfrak{g}$ (2.6 (17)). Let $g$ be the least positive rational integer in $\mathfrak{g}$, and $w_{\mathfrak{g}}$ the number of roots of unity congruent to 1 mod $\mathfrak{g}$ in $K$. Let $L_{\infty,\mathfrak{g}}(\chi, s) = (2\pi)^{-s}\Gamma(s)L(\chi, s)$, as in 1.1 (2). Then*

$$(1) \qquad L_{\infty,\mathfrak{g}}(\chi, 0) = \frac{-1}{12g \cdot w_{\mathfrak{g}}} \sum_{C \in Cl(\mathfrak{g})} \chi(C) \cdot \log |\varphi_{\mathfrak{g}}(C)|^2.$$

In this section we prove a $p$-adic analogue of this classical result. See also [K1], 10.4.

**5.2 Theorem.** *Notation and assumptions as above, let $p$ be a prime that splits in $K$, and write $\mathfrak{g} = \mathfrak{f}\mathfrak{p}^n$, $\mathfrak{p} \nmid \mathfrak{f}$. Define $L_{p,\mathfrak{f}}(\chi)$ as in 4.16 (49), and let $L_{p,\mathfrak{g}}(\chi) = L_{p,\mathfrak{f}}(\chi)(1 - \chi(\mathfrak{p}))$ if $n > 0$. If $\chi = 1$, assume $\mathfrak{f} \neq (1)$. Then*

$$(2) \quad L_{p,\mathfrak{g}}(\chi) = \frac{-1}{12gw_{\mathfrak{g}}} \cdot G(\chi^{-1}) \left(1 - \frac{\chi^{-1}(\mathfrak{p})}{p}\right) \cdot \sum_{C \in Cl(\mathfrak{g})} \chi(C) \cdot \log \varphi_{\mathfrak{g}}(C).$$

*Here "log" denotes any branch of the $p-adic$ logarithm ([Iw2] 4.1), and $G(\chi)$ is the Gauss sum 4.11 (30).*

Compare 4.12 (31). Forgetting for the moment the difference between complex and $p$-adic logarithms, we may say that (31) "extends" to $k = 0$. Recall that $L_{p,\mathfrak{f}}(\chi)$ is the integral of $\chi^{-1}$ (and not of $\chi$) against $\mu(\mathfrak{f})$.

PROOF: All the references below are to formulas from §4. Assume first that $w_\mathfrak{f} = 1$ and $n \geq 0$ is the exact power of $\mathfrak{p}$ in $\mathfrak{f}_\chi$ (but $\mathfrak{f}$ may be divisible by primes where $\chi$ is unramified). Then formulas (16) and (21) imply

$$\int_{\mathcal{G}} \chi^{-1}(\sigma) d\mu_\mathfrak{a}(\sigma) = \tau(\chi^{-1}) \left(1 - \frac{\chi^{-1}(\mathfrak{p})}{p}\right) \cdot \sum_{\mathfrak{c}} \chi(\mathfrak{c}) \delta_{0,n}(\sigma_\mathfrak{c}(\beta(\mathfrak{a})))^0$$

$$(3) \qquad = \tau(\chi^{-1}) \left(1 - \frac{\chi^{-1}(\mathfrak{p})}{p}\right) \sum \chi(\mathfrak{c}) \log(g_{\sigma_\mathfrak{c}(\beta(\mathfrak{a}))} \circ \theta)(\zeta_n - 1)$$

$$= \tau(\chi^{-1}) \left(1 - \frac{\chi^{-1}(\mathfrak{p})}{p}\right) \sum \chi(\mathfrak{c}) \cdot \log \sigma_\mathfrak{c}(e_n(\mathfrak{a}))^{\sigma_\mathfrak{q}}$$

where $e_n(\mathfrak{a}) = \Theta(\Omega; \mathfrak{p}^n \mathfrak{f} \Omega, \mathfrak{a})$ (see (23)), and $\mathfrak{q}$ is as in lemma 4.10: $(\mathfrak{q}, K(\mathfrak{f})/K) = \phi^n$, $\mathbf{N}\mathfrak{q} \equiv 1 \ mod\ p^n$. Note that $G(\chi^{-1}) = \chi^{-1}(\mathfrak{q})\tau(\chi^{-1})$.

Combining (3) with the definition of $L_{p,\mathfrak{g}}(\chi)$ and the fact that $\mu_\mathfrak{a} = \mu(\mathfrak{f})12(\sigma_\mathfrak{a} - \mathbf{N}\mathfrak{a})$, we get

(4)

$$L_{p,\mathfrak{g}}(\chi) \cdot 12 \cdot (\chi^{-1}(\mathfrak{a}) - \mathbf{N}\mathfrak{a}) = G(\chi^{-1}) \left(1 - \frac{\chi^{-1}(\mathfrak{p})}{p}\right) \sum \chi(\mathfrak{c}) \cdot \log \sigma_\mathfrak{c}(e_n(\mathfrak{a})).$$

Formula (2) follows from the last one, because by 2.4 (18)

$$e_n(\mathfrak{a})^g = \Theta(1; \mathfrak{g}, \mathfrak{a})^g = \varphi_\mathfrak{g}(1)^{\mathbf{N}\mathfrak{a} - \sigma_\mathfrak{a}},$$

and we have assumed $w_\mathfrak{g} = 1$.

The general case follows easily from the special one above, because if we denote the right hand side of (2) temporarily by $M_\mathfrak{g}(\chi)$, the distribution relation 2.3 (see also [K-La] p. 242) implies that for any prime $\mathfrak{l}$,

$$M_\mathfrak{g}^\mathfrak{l}(\chi) = \begin{cases} M_\mathfrak{g}(\chi) & \text{if } \mathfrak{l} | \mathfrak{g} \\ (1 - \chi(\mathfrak{l})) M_\mathfrak{g}(\chi) & \text{otherwise.} \end{cases}$$

The same relation holds between $L_{p,\mathfrak{g}}(\chi)$ and $L_{p,\mathfrak{g}^\mathfrak{l}}(\chi)$, as already mentioned in theorem 4.12(ii). We may therefore assume that $w_\mathfrak{f} = 1$ without any loss of generality.

**5.3 Corollary**. *$L_{p,(1)}$, the p – adic L function with conductor 1, has a true pole at the trivial character.*

PROOF: Fix an auxiliary *prime* ideal $\mathfrak{f}$. According to 4.12(iii) it is enough to show $L_{p,\mathfrak{f}}(1) = \mu(\mathcal{G}(\mathfrak{f})) \neq 0$ (note that if $\mathfrak{f}$ is divisible by two distinct primes this *is* zero). But (4) implies

$$(5) \qquad 12 \cdot \mu(\mathcal{G}(\mathfrak{f})) \cdot (1 - \mathbf{N}\mathfrak{a}) = \left(1 - \frac{1}{p}\right) \cdot \log N_{K(\mathfrak{f})/K}\Theta(1; \mathfrak{f}, \mathfrak{a})$$

and $N_{K(\mathfrak{f})/K}\Theta(1;\mathfrak{f},\mathfrak{a})$ is a $p$-adic unit which is not a global unit, since its valuation at $\mathfrak{f}$ is non-zero. In particular its $p$-adic logarithm is non-zero, as desired.

## 6. THE FUNCTIONAL EQUATION

The "two-variable" $p$-adic $L$ function satisfies a functional equation similar to the classical functional equation. In this respect the elliptic theory differs significantly from the cyclotomic theory of Kubota-Leopoldt $L$ functions. The reason is, vaguely speaking, that complex conjugation "acts on everything" and enables us to define an involution $\varepsilon \mapsto \check{\varepsilon}$ on characters of type $A_0$ of $K$, that preserves the critical ones (see also 6.7 below).

**6.1** For any grossencharacter $\varepsilon$ of $K$, let

$$(1) \qquad \check{\varepsilon}(\mathfrak{a}) = \varepsilon^{-1}(\overline{\mathfrak{a}})\mathbf{N}\mathfrak{a}^{-1}.$$

Clearly $\mathfrak{f}_{\check{\varepsilon}} = \overline{\mathfrak{f}}_{\varepsilon}$, and if the infinity type of $\varepsilon$ is $(k, j)$, that of $\check{\varepsilon}$ is $(-j-1, -k-1)$. In diagram 1.1(4) the infinity types of $\varepsilon$ and $\check{\varepsilon}$ are symmetrical about the line $k + j = -1$, so $\check{\varepsilon}$ is critical if and only if $\varepsilon$ is. Now let $\mathfrak{f}$ be an integral ideal, $(\mathfrak{f}, p) = 1$, and suppose $\mathfrak{f}_{\varepsilon} \mid \mathfrak{f}p^{\infty}$. Let $\mathcal{G} = Gal(K(\mathfrak{f}p^{\infty})/K)$ and $\check{\mathcal{G}} = Gal(K(\overline{\mathfrak{f}}p^{\infty})/K)$. If we consider $\varepsilon$ as a $p$-adic character of $\mathcal{G}$ (see 1.1(5)), and denote complex conjugation by $\rho$, $\check{\varepsilon}$ may be considered a character of $\check{\mathcal{G}}$, and (1) reads

$$(2) \qquad \check{\varepsilon}(\sigma) = \varepsilon^{-1}\mathbf{N}^{-1}(\rho\sigma\rho^{-1}).$$

As $\varepsilon\overline{\varepsilon} = \mathbf{N}^{k+j}$, we also have $L(\check{\varepsilon},0) = L(\varepsilon^{-1}\mathbf{N}^{-1},0) = L(\overline{\varepsilon},1+k+j)$, so the complex functional equation 1.1(3) becomes

$$(3) \qquad L_{\infty}(\varepsilon,0) = W \cdot (d_K \mathbf{N}\mathfrak{f}_{\varepsilon})^{\frac{1+k+j}{2}} \cdot L_{\infty}(\check{\varepsilon},0).$$

Artin's root number $W = W(\varepsilon)$ is a product of local factors $W = \prod W_v$ (over all places of $K$). Here $W_{\infty} = i^{|j-k|}$, and the finite factors $W_{\mathfrak{q}} = W_{\mathfrak{q}}(\varepsilon)$ are given as follows. Let $\varepsilon_0$ be the quasi-character on $K_{\mathbf{A}}^{\times}$ associated to $\varepsilon$ as in Tate's thesis. Thus $\varepsilon_0$ is trivial on principal ideles, and $\varepsilon_0(t_{\mathfrak{q}}) = \varepsilon(\mathfrak{q})$ if $(\mathfrak{q},\mathfrak{f}_{\varepsilon}) = 1$, and $t_{\mathfrak{q}}$ is any idele whose $\mathfrak{q}$-component is a uniformizer of $K_{\mathfrak{q}}$, and which is 1 at the other coordinates. Let $\mathfrak{q}^e \| (\sqrt{-d_K})$ and $\mathfrak{q}^m \| \mathfrak{f}_{\varepsilon}$, and denote by $tr$ the map

$$tr: \ K_{\mathfrak{q}} \xrightarrow{trace} \mathbf{Q}_q \rightarrow \mathbf{Q}_q/\mathbf{Z}_q \hookrightarrow \mathbf{Q}/\mathbf{Z}.$$

In the following sum, $u$ ranges over ideles trivial outside $\mathfrak{q}$, whose $\mathfrak{q}$-components represent $\mathcal{O}_{\mathfrak{q}}^{\times} \ mod \ 1 + \mathfrak{q}^m\mathcal{O}_{\mathfrak{q}}$. We then have ([La2] XIV, §8)

$$(4) \quad W_{\mathfrak{q}}(\varepsilon)^{-1} = \mathbf{N}\mathfrak{q}^{\frac{m(k+j-1)+e(k+j)}{2}} \cdot \sum_{u} \varepsilon_0(ut_{\mathfrak{q}}^{-m-e}) exp\left[2\pi i tr(ut_{\mathfrak{q}}^{-m-e})\right],$$

and this expression is independent of the choice of $t_{\mathfrak{q}}$ and $\{u\}$.

**6.2** For the functional equation we shall need to know explicitly the $\varsigma_n$ defined by 4.13(35). Write $\sqrt{-d_K}$ for $i\sqrt{d_K}$.

**Lemma.** *Let $\delta \in \mathbf{Z}_p^{\times}$ be the image of $\sqrt{-d_K}$ under $K \hookrightarrow K_{\mathfrak{p}} = \mathbf{Q}_p$. Then*

$$(5) \qquad \varsigma_n^{\delta} = e^{-2\pi i/p^n}.$$

PROOF: Write $\mathcal{O}_K = \mathbf{Z} + \mathbf{Z}\tau$, $\tau = \sqrt{-d_K}/2$ or $(1 + \sqrt{-d_K})/2$, according to whether $d_K = 0, 3 \ mod \ 4$. On the $p^n$-division points of the lattice $p^n\mathcal{O}_K$ we then have the following formula, giving the Weil pairing:

$$(6) \qquad e_{p^n}(x,y) = exp\left(2\pi i \frac{\overline{x}y - x\overline{y}}{p^n\sqrt{-d_K}}\right), \qquad x,y \in \mathcal{O}_K/p^n\mathcal{O}_K.$$

Now suppose $w_n$, $u_n$ are as in 4.13(15), i.e. $w_n \in \mathfrak{p}^n$, $u_n \in \overline{\mathfrak{p}}^n$, $w_n - u_n \equiv 1 \ mod\ p^n$. Then (6) gives us

$$\varsigma_n = e_{p^n}(w_n, u_n) = exp\left(2\pi i \frac{w_n - \overline{w}_n}{p^n\sqrt{-d_K}}\right).$$

Now $w_n - \overline{w}_n = (w_n - u_n) + (u_n - \overline{w}_n) \equiv 1 \ mod\ \overline{\mathfrak{p}}^n$, so also $w_n - \overline{w}_n \equiv -1 \ mod\ \mathfrak{p}^n$. It follows that in $K_{\mathfrak{p}}$ $(w_n - \overline{w}_n)/\sqrt{-d_K} \equiv -\delta^{-1} \ mod\ \mathfrak{p}^n$. Since $(w_n - \overline{w}_n)/\sqrt{-d_K} \in \mathbf{Q}$, the congruence holds $mod\ p^n$ if we consider $\delta$ in $\mathbf{Z}_p^x$. This proves the lemma.

**6.3** The key to the $p$-adic functional equation is a comparison between the local root numbers $W_{\mathfrak{p}}$ and $W_{\overline{\mathfrak{p}}}$, and the quantities denoted by $G(\varepsilon)$ in 4.14(37).

In the following, let $\varepsilon$ be a grossencharacter of type $(k, j)$, and $\mathfrak{f}_\varepsilon = \mathfrak{f}\mathfrak{p}^n\overline{\mathfrak{p}}^m$, $(\mathfrak{f}, p) = 1$. Let

$$\sigma_\delta \in Gal(K(\mathfrak{f}p^\infty)/K(\mathfrak{f}\overline{\mathfrak{p}}^\infty)) \tag{7}$$

be defined by $\sigma_\delta(\varsigma) = \varsigma^\delta$ for all $p$-power roots of unity $\varsigma$. Thus $\overline{\varphi}(\sigma_\delta) = 1$, $\varphi(\sigma_\delta) = \mathbf{N}(\sigma_\delta) = \delta$.

**Lemma.** *(i)* $W_{\mathfrak{p}}(\varepsilon)^{-1} = p^{\frac{\mathrm{n}}{2}(k+j+1)}\delta^{-k}\varepsilon(\sigma_\delta) \cdot G(\varepsilon^{-1})$

*(ii)* $W_{\overline{\mathfrak{p}}}(\varepsilon) = p^{-\frac{\mathrm{m}}{2}(k+j+1)}(-\delta)^{j+1}\check{\varepsilon}(\sigma_{-\delta}) \cdot G(\check{\varepsilon}^{-1})$.

PROOF: (i) When $\mathfrak{q} = \mathfrak{p}$ in formula (4), we have $e = 0$, and for $t_{\mathfrak{p}}$ we take $p = (\ldots 1, p, 1, \ldots) \in K_{\mathbf{A}}^x$. Then

$$\begin{aligned} W_{\mathfrak{p}}(\varepsilon)^{-1} &= p^{\frac{\mathrm{n}}{2}(k+j-1)} \sum_u \varepsilon_0^{-1}(p^n u) exp(2\pi i u^{-1} p^{-n}) \\ &= p^{\frac{\mathrm{n}}{2}(k+j-1)} . \varphi^{-k}\overline{\varphi}^{-j}(\mathfrak{p}^n) \cdot \sum_u \chi_0^{-1}(p^n u) exp(2\pi i u^{-1} p^{-n}), \end{aligned}$$

where we have written, as usual, $\varepsilon = \varphi^k\overline{\varphi}^j\chi$ with $\varphi$ of type (1,0) and conductor dividing $\mathfrak{f}\overline{\mathfrak{p}}^\infty$. Replace $exp(2\pi i p^{-n})$ by $\varsigma_n^{-\delta}$ (see (5)), and use local class field theory to identify the ideles $p^n u$, $u \in (\mathcal{O}_{\mathfrak{p}}/\mathfrak{p}^n\mathcal{O}_{\mathfrak{p}})^x$, with their Artin symbols in $\mathcal{G} =$

$Gal(K(\mathfrak{f}p^\infty)/K)$. Put $F' = K(\mathfrak{f}\overline{\mathfrak{p}}^\infty)$ and $F'_n = K(\mathfrak{f}\overline{\mathfrak{p}}^\infty\mathfrak{p}^n)$. Then $p^n u$ correspond to $S = \{\sigma \in Gal(F'_n/K) \mid \sigma|F' = (\mathfrak{p}^n, F'/K)\}$ (cf. 4.11). We arrive at

$$\begin{aligned} W_{\mathfrak{p}}(\varepsilon)^{-1} &= p^{\frac{n}{2}(k+j-1)} \cdot \varphi^{-k}\overline{\varphi}^{-j}(\mathfrak{p}^n) \cdot \sum_{\sigma \in S} \chi^{-1}(\sigma)\varsigma_n^{-\sigma\sigma_\delta} \\ &= p^{\frac{n}{2}(k+j+1)} \cdot \chi(\sigma_\delta) \cdot G(\varepsilon^{-1}). \end{aligned}$$

To obtain the expression in (1) note that $\varphi(\sigma_\delta) = \delta$ and $\overline{\varphi}(\sigma_\delta) = 1$.

(ii) It is easy to see, by "transport of structure", that

$$W_{\overline{\mathfrak{p}}}(\varepsilon) = W_{\mathfrak{p}}(\check{\varepsilon}^{-1}\mathrm{N}^{-1}).$$

The character $\check{\varepsilon}^{-1}\mathrm{N}^{-1}$ is of type $(j,k)$, and the power of $\mathfrak{p}$ in its conductor is $m$. Applying part (i) to it we get

$$W_{\overline{\mathfrak{p}}}(\varepsilon)^{-1} = p^{\frac{m}{2}(k+j+1)}\delta^{-j-1}\check{\varepsilon}^{-1}(\sigma_\delta) \cdot G(\check{\varepsilon}\mathrm{N})$$

where, to be perfectly honest, $\sigma_\delta$ is defined as in (7), but with $\overline{\mathfrak{f}}$ instead of $\mathfrak{f}$. Let $\sigma_{-1}$ be defined in the same way, with $-1$ replacing $\delta$ : $\sigma_{-1}(\varsigma) = \varsigma^{-1}$. Now $G(\check{\varepsilon}\mathrm{N}) = p^m G(\check{\varepsilon})$, and from $G(\check{\varepsilon})G(\check{\varepsilon}^{-1}) = p^{-m}\check{\varepsilon}(\sigma_{-1})(-1)^{j+1}$ we easily deduce the formula given in (ii).

**6.4** THE $p$-ADIC FUNCTIONAL EQUATION: Let $\varepsilon$ be a $p$-adic continuous character of $\mathcal{G} = Gal(K(\mathfrak{f}p^\infty)/K)$, $(\mathfrak{f}, p) = 1$, and assume that $\varepsilon$ is ramified at all the primes dividing $\mathfrak{f}$. As mentioned above, $\check{\varepsilon}$ (given by (2)) is a character of $\check{\mathcal{G}} = Gal(K(\overline{\mathfrak{f}}p^\infty)/K)$. Let us write simply $L_p(\varepsilon)$ for the $p$-adic $L$ function of $\varepsilon$, with modulus $\mathfrak{f}\overline{\mathfrak{p}}^\infty$ (see 4.16), and similarly $L_p(\check{\varepsilon})$. Thus

$$(8) \qquad \begin{cases} L_p(\varepsilon) = \int_{\mathcal{G}} \varepsilon^{-1}(\sigma)d\mu(\mathfrak{f}\overline{\mathfrak{p}}^\infty;\sigma) \\ L_p(\check{\varepsilon}) = \int_{\check{\mathcal{G}}} \check{\varepsilon}^{-1}(\sigma)d\mu(\overline{\mathfrak{f}}\overline{\mathfrak{p}}^\infty;\sigma). \end{cases}$$

**Theorem** (CF [K2] 5.3.7). *(i) There exists a p-adic unit $W^{padic}(\varepsilon)$ (the p-adic root number) for which*

$$(9) \qquad L_p(\varepsilon) = W^{padic}(\varepsilon) \cdot \frac{\check{\varepsilon}(\sigma_{-\delta})}{\varepsilon(\sigma_\delta)} \cdot L_p(\check{\varepsilon}).$$

*(ii) Suppose $\varepsilon$ is a grossencharacter of type $A_0$, of conductor $\mathfrak{f}\mathfrak{p}^n\overline{\mathfrak{p}}^m$. Then*

$$(10) \qquad W^{padic}(\varepsilon) = -i \prod_{\mathfrak{q}} W_{\mathfrak{q}}(\varepsilon)$$

*where the product is over all finite primes $\mathfrak{q}$, $(\mathfrak{q}, p) = 1$.*

PROOF: Let us assume first that $\varepsilon$ is a grossencharacter of type $(k, -k-1)$, $k < 0$, so its infinity type lies on the anticyclotomic line. Then $\check{\varepsilon}$ has the same infinity type, and both $\varepsilon$ and $\check{\varepsilon}$ lie within the interpolation range of theorem 4.14. Since $(1 - \frac{\varepsilon^{-1}(\mathfrak{p})}{p}) = (1 - \check{\varepsilon}(\overline{\mathfrak{p}}))$, and $\mathfrak{f}$ is the exact non-$p$ part of $\mathfrak{f}_\varepsilon$, formula 4.14(36) (with $\varepsilon^{-1}$ instead of $\varepsilon$) yields

$$(11) \quad \Omega_p^{2k+1} L_p(\varepsilon) = \Omega^{2k+1} \left(\frac{\sqrt{d_K}}{2\pi}\right)^{k+1} G(\varepsilon^{-1})(1 - \check{\varepsilon}(\overline{\mathfrak{p}}))(1 - \varepsilon(\overline{\mathfrak{p}}))\cdot L_\infty(\varepsilon, 0).$$

Similarly,

$$(12) \quad \Omega_p^{2k+1} L_p(\check{\varepsilon}) = \Omega^{2k+1} \left(\frac{\sqrt{d_K}}{2\pi}\right)^{k+1} G(\check{\varepsilon}^{-1})(1 - \varepsilon(\overline{\mathfrak{p}}))(1 - \check{\varepsilon}(\overline{\mathfrak{p}}))\cdot L_\infty(\check{\varepsilon}, 0).$$

Dividing (11) by (12), we deduce from the *complex* functional equation (3) that

$$\begin{aligned} \frac{L_p(\varepsilon)}{L_p(\check{\varepsilon})} &= \frac{G(\varepsilon^{-1})\cdot W(\varepsilon)}{G(\check{\varepsilon}^{-1})} \\ &= \delta^k \varepsilon^{-1}(\sigma_\delta)\cdot(-\delta)^{-k} \check{\varepsilon}(\sigma_{-\delta})\cdot W_\infty(\varepsilon)\cdot \prod_{\mathfrak{q}} W_{\mathfrak{q}}(\varepsilon), \end{aligned}$$

where we have used lemma 6.3. Finally, $W_\infty(\varepsilon) = i^{-1-2k}$ gives (9) and (10) in this case, and it is well known that (10) is a $p$-adic unit.

To treat the general case, let $K_\infty$ be the maximal $\mathbf{Z}_p^2$ extension of $K$, $G = Gal(K_\infty/K)$, $H = Gal(K(\mathfrak{f}p^\infty)/K_\infty)$, and $\check{H} = Gal(K(\overline{\mathfrak{f}}p^\infty)/K_\infty)$. Fix splittings $\mathcal{G} \cong G \times H$, $\check{\mathcal{G}} \cong G \times \check{H}$, which are compatible with the isomorphism $\mathcal{G} \cong \check{\mathcal{G}}$, $\sigma \mapsto \rho\sigma\rho^{-1}$ ($\rho$ is complex conjugation). Any character $\varepsilon$ of $\mathcal{G}$ can be written uniquely as $\varepsilon = \varepsilon_G \varepsilon_H$, where $\varepsilon_G$ is trivial on $H$, and vice versa. Furthermore $\varepsilon_G$ is ramified only at $\mathfrak{p}$ and $\overline{\mathfrak{p}}$, and if $\varepsilon$ is a grossencharacter of type $(k, j)$, so is $\varepsilon_G$, because $\varepsilon_H$ is of finite order.

Consider the two measures on $\mathcal{G}$ defined by

$$(13)\qquad \begin{cases} d\nu(\sigma) &= d\mu(\mathfrak{f}\bar{\mathfrak{p}}^{\infty}, \sigma\sigma_\delta) \\ d\check{\nu}(\sigma) &= \mathbf{N}^{-1}(\sigma)\cdot d\mu(\bar{\mathfrak{f}}\bar{\mathfrak{p}}^{\infty}, \rho\sigma^{-1}\rho^{-1}\sigma_{-\delta}). \end{cases}$$

It is easy to see that (9) is equivalent to the statement that if $\varepsilon_H$ is ramified at all the primes dividing $\mathfrak{f}$,

$$(14)\qquad \int_{\mathcal{G}} \varepsilon^{-1}(\sigma)d\nu(\sigma) = W^{padic}(\varepsilon)\cdot \int_{\mathcal{G}} \varepsilon^{-1}(\sigma)d\check{\nu}(\sigma).$$

Now fix $\varepsilon_H$ and vary $\varepsilon_G$. Observe that $W^{padic}(\varepsilon)$ depends only on $\varepsilon_H$, so it is unchanged. Thus (14) amounts to equality between *measures* on $G$:

$$(15)\qquad \varepsilon_H^{-1}(\nu) = W^{padic}(\varepsilon_H)\cdot\varepsilon_H^{-1}(\check{\nu}),$$

where we used $\varepsilon_H^{-1}$ to project $\mathbf{D}[[\mathcal{G}]]$ to $\mathbf{D}[[G]]$. We have verified before that for any grossencharacter $\varepsilon_G$ of type $(k, -k-1)$, $k < 0$, the integrals of $\varepsilon_G^{-1}$ against both sides of (15) coincide. Since we may twist $\varepsilon_G$ by any character of finite order without affecting its infinity type, there are enough admissible $\varepsilon_G$ to separate points in $\mathbf{D}[[G]]$. Thus (15) is proven, and with it (14) and (9). Formula (10) also follows from the above, as well as the fact that $W^{padic}(\varepsilon)$ is a unit.

REMARK: The involutive nature of $\varepsilon \mapsto \check{\varepsilon}$ together with (9) imply

$$(16)\qquad W^{padic}(\varepsilon)\cdot W^{padic}(\check{\varepsilon})\cdot\varepsilon\check{\varepsilon}(\sigma_{-1}) = 1.$$

This may be verified directly. For example, let us check it when $\mathfrak{f} = (1)$. Then (10) gives $W^{padic}(\varepsilon)\cdot W^{padic}(\check{\varepsilon}) = -1$. On the other hand $\varepsilon\check{\varepsilon}(\sigma_{-1}) = -\varepsilon(\sigma_{-1}\rho\sigma_{-1}^{-1}\rho^{-1}) = -1$, because $\sigma_{-1}\rho\sigma_{-1}^{-1}\rho^{-1}$ is the Artin symbol of the idele $\alpha = (\alpha_v)$ defined by $\alpha_{\mathfrak{p}} = \alpha_{\bar{\mathfrak{p}}} = -1$, $\alpha_v = 1$ otherwise, and we may change the remaining 1's to $-1$ because $\varepsilon$ is unramified outside $p$, thus obtaining a principal idele, whose Artin symbol is trivial.

**6.5** Of special interest are the *anticyclotomic characters* of $K$. These are, by definition, the $p$-adic characters satisfying

$$(17)\qquad \varepsilon = \check{\varepsilon}.$$

For such an $\varepsilon$ the sign in the functional equation (9) is

$$sgn(\varepsilon) = W^{padic}(\varepsilon)\cdot\varepsilon(\sigma_{-1}) = \pm 1, \tag{18}$$

because of (16). If $sgn(\varepsilon) = -1$, $\varepsilon$ is a zero of the $p$-adic $L$ function.

REMARK: An anticyclotomic character $\varepsilon$ must be ramified at some prime not above $p$, and, of course, its conductor is stable under complex conjugation. To see this compare (18) with (10). Alternatively, suppose a grossencharacter $\varepsilon$ satisfies $\varepsilon = \check{\varepsilon}$. Then if $\gamma_v$ denotes the idele which is $-1$ at $v$ and 1 elsewhere, $\varepsilon(\gamma_{\bar{\mathfrak{p}}}) = \check{\varepsilon}(\gamma_{\mathfrak{p}}) = \varepsilon(\gamma_{\mathfrak{p}})$, hence $\varepsilon(\gamma_\infty\gamma_{\mathfrak{p}}\gamma_{\bar{\mathfrak{p}}}) = \varepsilon(\gamma_\infty) = -1$, because the infinity type of $\varepsilon$ is $(k, -k-1)$. But if $\varepsilon$ is unramified outside $p$, $\varepsilon(\gamma_\infty\gamma_{\mathfrak{p}}\gamma_{\bar{\mathfrak{p}}}) = \varepsilon(-1,-1,\ldots) = 1$. Our argument actually proves that $\varepsilon$ must be ramified at a non-split prime.

**6.6** The functional equation has the following obvious, but important, corollary.

**Corollary.** $L_p(\varepsilon) = 0 \Leftrightarrow L_p(\check{\varepsilon}) = 0.$

**6.7** Another corollary of the functional equation is the following.

**Corollary.** *Notation as in theorem 4.14, the conclusion of that theorem (the interpolation formula (36)) is valid for any grossencharacter* $\varepsilon$ *of type* $(k, j)$, *where* $k > 0$ *and* $j \leq 0$.

In other words, the interpolation range is extended to half of the critical values. We cannot extend this result to the other half. The *tautological* functional equation

$$L(\varepsilon, s) = L(\varepsilon\circ\rho, s), \tag{19}$$

where $\rho$ is complex conjugation, *does not* have a $p$-adic analogue. Note that the $p$-adic functional equation was the counterpart of (3), which is the "composition" of (19) with the classical functional equation 1.1(3), and not of either of these two alone.

# CHAPTER III
# APPLICATIONS TO CLASS FIELD THEORY

There are two main themes to Iwasawa theory of elliptic curves with complex multiplication, the interplay between which enriches our understanding of both. One deals with abelian extensions of quadratic imaginary fields, especially their units and ideal class groups. The other is concerned with the arithmetic of CM elliptic curves, and problems having their origin in diophantine equations. Although these two themes are intricately woven, we shall try to concentrate on the first here, and postpone the striking applications to the conjecture of Birch and Swinnerton-Dyer to the fourth and last chapter.

To put things in perspective, let us regress a little and discuss cyclotomic Iwasawa theory. Motivated by Weil's work on curves over finite fields, Iwasawa sought an analogue for the Jacobian variety in the case of number fields. Unable to find a satisfactory answer in the large, he restricted his attention to the $p$-primary part of the Jacobian ($p$ not equal to the characteristic). Consider $\mathbf{Q}$ as a base field. The proposed analogue of the "geometric" curve (over the algebraic closure of the finite field) is $F_\infty = \mathbf{Q}(p^\infty)$, the maximal abelian extension of $\mathbf{Q}$ unramified outside $p$ (the real subfield of $\mathbf{Q}(\mu_{p^\infty})$). The analogue of the $p$-adic Tate module of the Jacobian would be the *Iwasawa module* $\mathcal{X} = Gal(M_\infty/F_\infty)$, where $M_\infty$ is the maximal pro-$p$ abelian extension of $F_\infty$ unramified outside $p$. The study of $\mathcal{X}$ as a module over $\Lambda = \mathbf{Z}_p[[\mathcal{G}]]$, $\mathcal{G} = Gal(F_\infty/\mathbf{Q})$, is the focal point of Iwasawa theory.

In Weil's theory, the zeta function of the curve is essentially given by the characteristic polynomial of Frobenius on the $p$-adic Tate module of the Jacobian. Its zeroes are the reciprocals of the eigenvalues of Frobenius in its action on the Tate module. The celebrated theorem of Mazur and Wiles (formerly the "*main conjecture*" of Iwasawa) asserts that the eigencharacters by which $\mathcal{G}$ acts on $\mathcal{X}$ (a free $\mathbf{Z}_p$ module of finite rank) are precisely the reciprocals of the zeroes of the

Kubota-Leopoldt $p$-adic $L$ function. Note how elegant this formulation becomes when we view the $p$-adic $L$ function as a function of *characters* on $\mathcal{G}$ (see the discussion preceding I.3). Although the semi-simplicity of $\mathcal{X}$ as a $\Lambda$-module remains unresolved, the theorem of Mazur and Wiles supplies the link between the $p$-adic analytic side and the algebro-arithmetic theory of $\mathbf{Z}_p$ extensions.

Our main aim here is to formulate a corresponding main conjecture in the elliptic case. See 1.9-1.11. We shall give some evidence in favour of it, proving the equality of the Iwasawa invariants of the two modules which are supposed to have the same characteristic polynomial. Thus instead of matching the roots of two polynomials, we show that their degrees are equal. We also discuss without proofs Gillard's theorem on the vanishing of the $\mu$-invariant, the important recent work of K. Rubin on relations in ideal class groups of abelian extensions of $K$, and its relation to the main conjecture.

Throughout chapter III we assume $p > 2$.

## 1. THE MAIN CONJECTURE

**1.1** We keep the notation of chapter II. Thus $K$ is our quadratic imaginary field. Fix an integral ideal $\mathfrak{f}$ of $K$, and a split prime $\mathfrak{p}$, $(\mathfrak{p}, \mathfrak{f}) = 1$. Let $F = K(\mathfrak{f})$ and $F_n = K(\mathfrak{f}\mathfrak{p}^n)$ be the corresponding ray class fields. Every prime $\mathfrak{P}$ of $F$ lying above $\mathfrak{p}$ is totally ramified in $F_\infty$.

Denote the integers of the completion of $\mathbf{Q}_p^{ur}$ by $\mathbf{D}$, let $\mathcal{G} = \mathcal{G}(\mathfrak{f}) = Gal(K(\mathfrak{f}\mathfrak{p}^\infty)/K)$, and $\Lambda = \Lambda(\mathcal{G}, \mathbf{D})$, the convolution algebra of $\mathbf{D}$-valued measures on $\mathcal{G}$. Recall (II.4.1) that

$$(1) \qquad \mathcal{U} = \mathcal{U}(\mathfrak{f}) = \varprojlim U_n$$

denoted the inverse limit of the groups $U_n$ of (semi-local) principal units in the completion of $F_n$ at $\mathfrak{p}$. Unlike II.4.1, we do not assume anymore $w_\mathfrak{f} = 1$. In particular, $\mathfrak{f}$ may be trivial. $\mathcal{U}$ is a torsion-free pro-$p$ group, and also a $\mathbf{Z}_p[[\mathcal{G}]]$-module.

Let $Z$ be the decomposition group of $\mathfrak{p}$ in $\mathcal{G}$, $G = Gal(F_\infty/F)$ and $\Gamma = Gal(F_\infty/F_1)$ so that

$$K - \cdot - F - F_1 - F_\infty$$

$$\mathcal{G} \supset Z \supset G \supset \Gamma \supset (1)$$

is the field diagram. Let $\mathfrak{P}$ be any prime of $F$ above $\mathfrak{p}$, and consider the groups $\mu_{p^\infty}(F_\infty)$ and $\mu_{p^\infty}(F_{\infty,\mathfrak{P}})$. Let their annihilators in $\mathbf{Z}_p[[\mathcal{G}]]$ and $\mathbf{Z}_p[[Z]]$ be denoted by $J_0$ and $J_1$ respectively, and define three ideals in $\Lambda$:

$$\Lambda_0 = J_0\Lambda$$

$$\Lambda_1 = J_1\Lambda$$

$$\Lambda_{00} = \Lambda_0 \cap (\text{the augmentation ideal of } \Lambda).$$

EXERCISE: (i) Show that $\Lambda_0$ is generated by $\sigma_\mathfrak{a} - \mathbf{N}\mathfrak{a}$, $(\mathfrak{a}, 6\mathfrak{f}\mathfrak{p}) = 1$, where $\sigma_\mathfrak{a} = (\mathfrak{a}, F_\infty/K)$.

(ii) Let $\mathfrak{P}$ be a prime of $F$ dividing $\mathfrak{p}$ and $p^N$ the number of $p$-power roots of unity in $F_{\infty,\mathfrak{P}}$. If $w_\mathfrak{f} = 1$, $\xi$ is the unique generator of $N_{F/K}\mathfrak{P}$ satisfying $\xi \equiv 1 \; mod \; \mathfrak{f}$, and $f = f(\mathfrak{P}/\mathfrak{p})$ the relative degree, show that $N$ is the precise power of $\mathfrak{p}$ in $p^f\xi^{-1} - 1$.

(iii) Show that $\Lambda/\Lambda_1 \cong \mathbf{D}/(p^N)[\mathcal{G}/Z]$.

**1.2** THE HOMOMORPHISM $i$: When $w_\mathfrak{f}$ (the number of roots of unity in $K$ congruent to 1 modulo $\mathfrak{f}$) is 1, $\mathcal{U}$ was embedded in $\Lambda$ by means of the $\mathcal{G}$-module homomorphism $i : \beta \mapsto \mu^0_\beta$ (II.4.6 and 4.7).

**Lemma.** *(i) The map $i$ is independent of the elliptic curve $E$ used in its construction. It is canonically associated to the extension $F_\infty/K$, and the choice of $(\varsigma_n)$.*

*(ii) Suppose $\mathfrak{f}|\mathfrak{g}$ and $w_\mathfrak{f} = 1$. Set $\Lambda(\mathfrak{g}) = \Lambda(\mathcal{G}(\mathfrak{g}), \mathbf{D})$, and let*

$$\pi_{\mathfrak{g},\mathfrak{f}} : \Lambda(\mathfrak{g}) \to \Lambda(\mathfrak{f})$$

$$\eta_{\mathfrak{f},\mathfrak{g}} : \Lambda(\mathfrak{f}) \to \Lambda(\mathfrak{g})$$

*be the maps corresponding to restriction and corestriction on the Galois groups. Then the following diagrams commute ($N_{\mathfrak{g},\mathfrak{f}}$ is the norm from $K(\mathfrak{g}\mathfrak{p}^\infty)$ to $K(\mathfrak{f}\mathfrak{p}^\infty)$)*

$$\text{(2)}\qquad \begin{array}{ccc} \mathcal{U}(\mathfrak{g}) & \xrightarrow{i(\mathfrak{g})} & \Lambda(\mathfrak{g}) \\ N_{\mathfrak{g},\mathfrak{f}}\downarrow & & \downarrow \pi_{\mathfrak{g},\mathfrak{f}} \\ \mathcal{U}(\mathfrak{f}) & \xrightarrow{i(\mathfrak{f})} & \Lambda(\mathfrak{f}) \end{array} \qquad \begin{array}{ccc} \mathcal{U}(\mathfrak{g}) & \xrightarrow{i(\mathfrak{g})} & \Lambda(\mathfrak{g}) \\ incl.\uparrow & & \uparrow \eta_{\mathfrak{f},\mathfrak{g}} \\ \mathcal{U}(\mathfrak{f}) & \xrightarrow{i(\mathfrak{f})} & \Lambda(\mathfrak{f}). \end{array}$$

(We remind the reader that for $\tau \in \mathcal{G}(\mathfrak{f})$, $\eta_{\mathfrak{f},\mathfrak{g}}(\tau) = \sum_{\sigma\mapsto\tau}\sigma$).

PROOF: (i) Obvious in view of II.4.6 (14).

(ii) Both parts follow from the definitions and are left to the reader.

**1.3 Proposition.** *There is a unique way to extend the definition of $i(\mathfrak{f})$ to all $\mathfrak{f}$ so that (2) remains commutative. Furthermore, $i$ induces an isomorphism*

$$\text{(3)}\qquad i: \mathcal{U}\,\hat{\otimes}_{\mathbf{Z}_p}\mathbf{D} \simeq \Lambda_1.$$

PROOF: Assume first that $w_{\mathfrak{f}} = 1$, so that $i$ is defined (II.4.6). Let $\mathcal{U}_z$ be the inverse limit of the groups of principal units in $F_{n,\mathfrak{P}}$, and $\Lambda_{1,z}$ the ideal of $\Lambda_z = \mathbf{D}[[Z]]$ generated by $J_1$ (see 1.1 for notation). Then $\mathcal{U} = Ind_Z^{\mathcal{G}}\,\mathcal{U}_z$ and $\Lambda_1 = Ind_Z^{\mathcal{G}}\,\Lambda_{1,z}$. In I.3.7 we proved that $i$ maps $\mathcal{U}_z\,\hat{\otimes}_{\mathbf{Z}_p}\mathbf{D}$ isomorphically onto $\Lambda_{1,z} = Ker(j)$. Our claim is the semi-local version of this and follows from it since $\Lambda$ is free over $\Lambda_z$.

In the general case one may use (2) to *define* $i(\mathfrak{f})$. Pick $\mathfrak{g}$ with $w_{\mathfrak{g}} = 1$, $\mathfrak{f}|\mathfrak{g}$. It is easy to check that $\pi_{\mathfrak{g},\mathfrak{f}}(i(\mathfrak{g})(\beta)) = 0$ if and only if $N_{\mathfrak{g},\mathfrak{f}}\beta = 1$. On the other hand $N_{\mathfrak{g},\mathfrak{f}}$ is surjective. To see this consider the norm from $K(\mathfrak{g}\mathfrak{p}^n)$ to $K(\mathfrak{f}\mathfrak{p}^n)$ in two steps: $K(\mathfrak{g}\mathfrak{p}^n) \to K(\mathfrak{g})K(\mathfrak{f}\mathfrak{p}^n) \to K(\mathfrak{f}\mathfrak{p}^n)$. The first step is ramified at $\mathfrak{p}$, but only tamely, since the degree (for large $n$) is $w_{\mathfrak{f}}$, and $(w_{\mathfrak{f}}, p) = 1$. The second step is unramified at $\mathfrak{p}$. In any case the norm on *principal* units is surjective. Thus $N_{\mathfrak{g},\mathfrak{f}}$ is surjective and there is a unique way to define $i(\mathfrak{f})$ so that (2) is still commutative. It also follows from what was said above that (3) remains valid.

**1.4** THE GROUP $\mathcal{C}_{\mathfrak{f}}$: Let $\mathfrak{f}$ be any integral ideal of $K$ and $C_n = C_{\mathfrak{f}\mathfrak{p}^n}$ the group of primitive Robert units of conductor $\mathfrak{f}\mathfrak{p}^n$, $n \geq 1$. Recall their definition (II.2.7). If $\mathfrak{f} \neq (1)$, then each $\Theta(1; \mathfrak{f}\mathfrak{p}^n, \mathfrak{a})$, $(\mathfrak{a}, 6\mathfrak{f}\mathfrak{p}) = 1$ is a $12^{th}$ power of a unit in $K(\mathfrak{f}\mathfrak{p}^n) = F_n$. Denote by $\Theta_{12}(1; \mathfrak{f}\mathfrak{p}^n, \mathfrak{a})$ one such root. The group $C_n$ is generated by $\Theta_{12}(1; \mathfrak{f}\mathfrak{p}^n, \mathfrak{a})$, $(\mathfrak{a}, 6\mathfrak{f}\mathfrak{p}) = 1$, and by the roots of unity in $F_n$. If $\mathfrak{f} = 1$, $\Theta(1; \mathfrak{p}^n, \mathfrak{a})$ still admits the twelfth root $\Theta_{12}(1; \mathfrak{p}^n, \mathfrak{a})$ in $F_n$, but these are not units. A product of the form $\prod \Theta_{12}(1; \mathfrak{p}^n, \mathfrak{a})^{m(\mathfrak{a})}$ $((\mathfrak{a}, 6\mathfrak{p}) = 1)$ is a unit if and only if $\sum m(\mathfrak{a})(\mathbf{N}\mathfrak{a} - 1) = 0$ (exercise II.2.4). We let $C_n$ be the group generated by all such products and by the roots of unity in $F_n$. In any case $C_n$ is Galois-stable, and $N_{m,n} C_m \equiv C_n$ modulo roots of unity (proposition II.2.5(i)). Notice that since we assume $p > 2$, and since $p$ is split in $K$, $w_{\mathfrak{p}} = 1$. Let $\overline{C}_n$ be the closure of $C_n$ in $U_n \times V_n$, $< \overline{C}_n >$ its projection to $U_n$, and

$$C_{\mathfrak{f}} = \varprojlim < \overline{C}_n > \subset \mathcal{U}(\mathfrak{f}). \tag{4}$$

**Proposition.** *The map $i : \mathcal{U}(\mathfrak{f}) \to \Lambda$ induces an isomorphism*

$$i : \mathcal{C}_{\mathfrak{f}} \hat{\otimes}_{\mathbf{Z}_p} \mathbf{D} \simeq \begin{cases} \mu(\mathfrak{f})\Lambda_0 & \mathfrak{f} \neq (1) \\ \mu(1)\Lambda_{00} & \mathfrak{f} = (1) \end{cases} \tag{5}$$

*where $\mu(\mathfrak{f})$ is the measure constructed in II.4.12.*

REMARKS: (i) If $\mathfrak{f} \neq (1)$, $\mathcal{C}_{\mathfrak{f}}$ may be described as the group generated by the $12^{th}$ roots of $\beta(\mathfrak{a})$ in $\mathcal{U}(\mathfrak{f})$, for $(\mathfrak{a}, 6\mathfrak{f}\mathfrak{p}) = 1$. If $(p, 6) = 1$ there is no need to extract $12^{th}$ roots. If $\mathfrak{g} = \mathfrak{f}\mathfrak{l}$ with $\mathfrak{l}$ prime, and $\mathfrak{f} \neq (1)$, proposition II.2.5 (i) implies that

$$N_{\mathfrak{g},\mathfrak{f}}\, \mathcal{C}_{\mathfrak{g}} = \begin{cases} \mathcal{C}_{\mathfrak{f}}^{1-\sigma_{\mathfrak{l}}^{-1}} & \text{if } \mathfrak{l} \nmid \mathfrak{f} \\ \mathcal{C}_{\mathfrak{f}} & \text{if } \mathfrak{l} \mid \mathfrak{f}. \end{cases} \tag{6}$$

Compare II.4.12 (32) and (5). If $\mathfrak{f} = (1)$, $N_{\mathfrak{l},(1)} \mathcal{C}_{\mathfrak{l}} \subset \mathcal{C}_{(1)}$.

(ii) $\mu(1)$ is only a pseudo-measure, but for any $\lambda$ in the augmentation ideal, $\mu(1) \cdot \lambda \in \Lambda$ (II.4.12 (iii)).

PROOF: Recall the notation of II.4 in case $w_{\mathfrak{f}} = 1$ : $i(\beta(\mathfrak{a})) = \mu_{\mathfrak{a}} = 12 \cdot \mu(\mathfrak{f}) \cdot (\sigma_{\mathfrak{a}} - \mathbf{N}\mathfrak{a})$. Our assertion follows from this, and from the definitions of $\mathcal{C}_{\mathfrak{f}}$

and $\Lambda_0$. In the rare event that $\mathfrak{f} \neq (1)$ but $w_\mathfrak{f} > 1$, replace $\mathfrak{f}$ by $\mathfrak{g} = \mathfrak{f}^m$ for some large $m$ and "push down" (5) from level $\mathfrak{g}$ to level $\mathfrak{f}$ using (6) and II.4.12 (32).

If $\mathfrak{f} = (1)$ the situation is truly different because now $e_n(\mathfrak{a}) = \Theta(1;\mathfrak{p}^n,\mathfrak{a}) \notin C_n$. Using exercise II.2.4 we deduce that $\Pi e_n(\mathfrak{a})^{m(\mathfrak{a})}$ is a unit if and only if $\sum m(\mathfrak{a})(\sigma_\mathfrak{a} - \mathbf{N}\mathfrak{a})$ belongs to the augmentation ideal, i.e. $\sum m(\mathfrak{a})(1-\mathbf{N}\mathfrak{a}) = 0$. This explains why in (5) we substituted $\Lambda_{00}$ for $\Lambda_0$. The rest of the argument is identical to the previous case ($\mathfrak{f} \neq (1)$).

**1.5 Corollary.** *The map $i$ induces an isomorphism*

$$\mathcal{U}(\mathfrak{f})/C_\mathfrak{f} \hat{\otimes}_{\mathbf{Z}_p} \mathbf{D} \cong \begin{cases} \Lambda_1/\mu(\mathfrak{f})\Lambda_0 & \mathfrak{f} \neq (1) \\ \Lambda_1/\mu(1)\Lambda_{00} & \mathfrak{f} = (1). \end{cases} \tag{7}$$

*In particular $\mathcal{U}(\mathfrak{f})/C_\mathfrak{f}$ is a torsion noetherian $\mathbf{Z}_p[[\mathcal{G}]]$-module (i.e. killed by a non-zero-divisor from the ring).*

A remarkable observation is that $\Lambda_1 \subset \Lambda_0$, so *unless* $\Lambda_1 = \Lambda_0$, $\mu(\mathfrak{f})$ is not a unit. In particular, if there are no $p$-power roots of unity in $F_\infty$, but there are some in $F_{\infty,\mathfrak{P}}$, $\mu(\mathfrak{f})$ is not invertible. Notice that in this case, in the language of I.3.7, any $E$ as in II.4.1 is *anomalous* above $\mathfrak{p}$ (meaning that $\hat{E}$ is anomalous).

Note also that when $\mathfrak{f} = (1)$ there are no non-trivial $p$-power roots of unity in $F_{\infty,\mathfrak{P}}$, so $\Lambda_1 = \Lambda_0 = \Lambda$, and $\Lambda_{00}$ is the augmentation ideal.

**1.6** The measure $\mu(\mathfrak{f})$ is deprived of the Euler factors at the primes dividing $\mathfrak{f}$ (II.4.12 (32)). While it is impossible to restore them to $\mu(\mathfrak{f})$, we may remedy that matter on the left hand side of (7), by considering a different module.

For any divisor $\mathfrak{g}$ of $\mathfrak{f}$, $C_\mathfrak{g} \subset \mathcal{U}(\mathfrak{g}) \subset \mathcal{U}(\mathfrak{f})$, and $N_{\mathfrak{f},\mathfrak{g}}C_\mathfrak{f} \subset C_\mathfrak{g}$, where $N_{\mathfrak{f},\mathfrak{g}}$ denotes the norm from $K(\mathfrak{f}\mathfrak{p}^\infty)$ to $K(\mathfrak{g}\mathfrak{p}^\infty)$, as before. In fact, equality prevails here if $\mathfrak{f}$ and $\mathfrak{g}$ are divisible by the same primes (6).

Definition. *The Iwasawa module of elliptic units in $K(\mathfrak{f}\mathfrak{p}^\infty)/K$ is*

$$C(\mathfrak{f}) = \prod_{\mathfrak{g}|\mathfrak{f}} C_\mathfrak{g}.$$

If $\mathfrak{f}$ is indeed the conductor of $K(\mathfrak{f})$, we refer to $C_\mathfrak{f}$ as the *primitive* (or new) part of $C(\mathfrak{f})$.

**1.7** THE FUNDAMENTAL EXACT SEQUENCE: Let $E_n$ be the global units in $F_n$, $\overline{E}_n$ their closure in $U_n \times V_n$, and $< \overline{E}_n >$ the projection of $\overline{E}_n$ to $U_n$. Let

$$(8) \qquad \mathcal{E}(\mathfrak{f}) = \varprojlim < \overline{E}_n > \subset \mathcal{U}(\mathfrak{f}).$$

We thus have a chain of inclusions

$$(9) \qquad \begin{cases} \overline{C}_n \subset \ldots \subset \overline{E}_n \subset U_n \times V_n \\ \mathcal{C}_\mathfrak{f} \subset \mathcal{C}(\mathfrak{f}) \subset \mathcal{E}(\mathfrak{f}) \subset \mathcal{U}(\mathfrak{f}). \end{cases}$$

Let $M_n$ be the *maximal abelian pro-p extension of* $F_n$ *unramified at (the primes above)* $\mathfrak{p}$. Let us put $(1 \leq n)$

$$(10) \qquad \mathcal{X}_n = Gal(M_n/F_n), \qquad \mathcal{X} = \mathcal{X}(\mathfrak{f}) = Gal(M_\infty/F_\infty).$$

The first is a $\mathbf{Z}_p[Gal(F_n/K)]$-module, and the second a topological $\mathbf{Z}_p[[\mathcal{G}]]$-module. Clearly $M_\infty = \cup M_n$.

Class field theory provides an exact sequence

$$(11) \qquad 0 \to U_n/ < \overline{E}_n > \xrightarrow{a} \mathcal{X}_n \to \mathcal{W}_n \to 0$$

where $\mathcal{W}_n = Gal(L_n/F_n)$ is the Galois group of the maximal *unramified* abelian $p$-extension of $F_n$ (the Hilbert $p$-class field). The injection $a$ is the (idelic) Artin symbol. Taking projective limits over $n$, (11) yields

$$(12) \qquad 0 \to \mathcal{U}(\mathfrak{f})/\mathcal{E}(\mathfrak{f}) \to \mathcal{X}(\mathfrak{f}) \to \mathcal{W}(\mathfrak{f}) \to 0.$$

We shall re-write (12) as a four-term exact sequence

$$(13) \qquad 0 \to \mathcal{E}(\mathfrak{f})/\mathcal{C}(\mathfrak{f}) \to \mathcal{U}(\mathfrak{f})/\mathcal{C}(\mathfrak{f}) \to \mathcal{X}(\mathfrak{f}) \to \mathcal{W}(\mathfrak{f}) \to 0.$$

The idea is that $\mathcal{U}/\mathcal{C}$ is somewhat better understood, at least in terms of $L$ functions, than $\mathcal{U}/\mathcal{E}$.

The "main conjecture" compares $\mathcal{U}(\mathfrak{f})/\mathcal{C}(\mathfrak{f})$ with $\mathcal{X}(\mathfrak{f})$ as representation spaces for $\mathcal{G}$. Considered as $\mathbf{Z}_p[[\mathcal{G}]]$-modules, both of them are torsion and noetherian. This was already verified for $\mathcal{U}/\mathcal{C}$, and it is a well known fact about $\mathcal{W}(\mathfrak{f})$, which holds in any $\mathbf{Z}_p$ extension ([Wa] 13.3). The exactness of (13) implies it for $\mathcal{X}(\mathfrak{f})$ too.

**1.8** Let $K_\infty$ be the unique $\mathbf{Z}_p$ extension of $K$ unramified outside $\mathfrak{p}$, $\Gamma' = Gal(K_\infty/K)$, $H' = Gal(F_\infty/K_\infty)$, and fix, once and for all, an identification

$$\mathcal{G} \simeq H' \times \Gamma'$$

so that characters of $H'$ will be naturally considered as characters of $\mathcal{G}$ too. If $K(1) \cap K_\infty = K_t$ ($[K_t : K] = p^t$), then the image of $\Gamma$ in $\Gamma'$, under restriction to $K_\infty$, is $\Gamma'^{p^t}$.

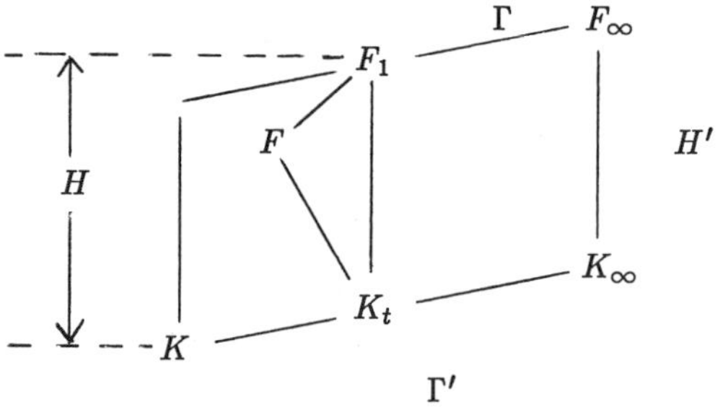


Let $\mathbf{D}'$ be the finite extension of $\mathbf{D}$ generated by the values of all $\chi$ from $\hat{H}'$. We want to decompose the modules in (13) according to $\chi \in \hat{H}'$, and study each piece as a $\mathbf{D}'[[\Gamma']]$-module. If $p \nmid [F : K]$, then $H = H'$, $\Gamma = \Gamma'$, $\mathbf{D} = \mathbf{D}'$ and any $\mathbf{D}[[\mathcal{G}]]$ module $M$ automatically breaks up as the direct sum of its $\chi$-components $M^\chi$. In general, let us agree to consider $\mathbf{D}'[[\Gamma']]$ as a $\Lambda = \mathbf{D}[[\mathcal{G}]]$-module via $\chi$, extended to a homomorphism of the group algebras. Put

$$M_\chi = M \otimes_{\Lambda,\chi} \mathbf{D}'[[\Gamma']], \tag{14}$$

the largest quotient of $M$ on which $H'$ acts through $\chi$ (scalars extended to $\mathbf{D}'$).

Recall the structure theory of $\mathbf{D}'[[\Gamma']]$-modules ([Wa] 13.2). Such a module is *pseudo-null* if it is of finite length. $M$ and $M'$ are *pseudo-isomorphic* if there exists an exact sequence

$$0 \to \varepsilon \to M \to M' \to \varepsilon' \to 0$$

where $\varepsilon$ and $\varepsilon'$ are pseudo-null. This is an equivalence relation if $M$ and $M'$ are noetherian and torsion. Every noetherian torsion module $M$ is pseudo isomorphic to an *elementary module*

$$E = \prod_{1 \le i \le r} \mathbf{D}'[[\Gamma']]/(f_i). \tag{15}$$

The ideal $(\prod f_i)$ is an invariant of $M$, called its ***characteristic ideal.*** Any generator of it is called a *characteristic power series* of $M$ (the terminology stems from the fact that $\mathbf{D}'[[\Gamma']] \cong \mathbf{D}'[[X]]$, cf. I.3.1). We denote the characteristic ideal by $char.(M)$.

If the $p$-torsion submodule of $M$ is pseudo-null, then $M_\chi$ is pseudo-isomorphic to $M^\chi$ = the largest submodule of $M$ on which $H'$ acts via $\chi$. We shall see later that this is the case with $\mathcal{X}(\mathfrak{f})$.

**1.9** THE MAIN CONJECTURE. *Let* $\mathfrak{f}$ *be any integral ideal and fix a decomposition* $\mathcal{G}(\mathfrak{f}) \simeq H' \times \Gamma'$ *as above. Then for any* $\chi \in \hat{H}'$,

$$char.\ (\mathcal{U}(\mathfrak{f})/\mathcal{C}(\mathfrak{f}))_\chi = char.\ \mathcal{X}(\mathfrak{f})_\chi. \tag{16}$$

**1.10** The following lemma is more or less obvious, in light of 1.5. Let $\mu(\mathfrak{f};\chi) = \chi(\mu(\mathfrak{f})) \in \mathbf{D}'[[\Gamma']]$.

**Lemma.** *Suppose* $\mathfrak{f}_\chi$ *is* $\mathfrak{g}$ *or* $\mathfrak{g}\mathfrak{p}$ $(\mathfrak{g}|\mathfrak{f})$*. Then*

$$char.\ (\mathcal{U}(\mathfrak{f})/\mathcal{C}(\mathfrak{f}))_\chi = \begin{cases} \mu(\mathfrak{g};\chi) & \chi \neq 1 \\ (\gamma_0 - 1)\mu(1;1) & \chi = 1, \end{cases} \tag{17}$$

*where* $\gamma_0$ *is a topological generator of* $\Gamma'$.

PROOF: Consider first the map $\chi \circ i(\mathfrak{f}) : \mathcal{U}(\mathfrak{f}) \hat{\otimes} \mathbf{D}' \to \mathbf{D}'[[\Gamma']]$. We claim that $\chi \circ i(\mathfrak{f})(\mathcal{C}(\mathfrak{f}) \hat{\otimes} \mathbf{D}') \subset \chi \circ i(\mathfrak{g})(\mathcal{C}_\mathfrak{g} \hat{\otimes} \mathbf{D}')$ (here we consider $\chi$ as a character on $\mathcal{G}(\mathfrak{f})$ and $\mathcal{G}(\mathfrak{g})$ simultaneously), and that the cokernel of this inclusion is pseudo-null. Indeed, $\chi \circ i(\mathfrak{g})(\mathcal{C}_\mathfrak{g}) = \chi \circ i(\mathfrak{g})(\mathcal{C}(\mathfrak{g}))$ because $\chi$ has conductor $\mathfrak{g}$ or $\mathfrak{g}\mathfrak{p}$. Also, proposition II.2.5 implies $N_{\mathfrak{f},\mathfrak{g}}\mathcal{C}(\mathfrak{f}) \subset \mathcal{C}(\mathfrak{g})$ (compare (6)—here we have to deal with $N_{\mathfrak{f},\mathfrak{g}}\mathcal{C}_\mathfrak{h}$ for any $\mathfrak{h}|\mathfrak{f}$). Thus, $\chi \circ i(\mathfrak{f})(\mathcal{C}(\mathfrak{f})) = \chi \circ \pi_{\mathfrak{f},\mathfrak{g}} \circ i(\mathfrak{f})(\mathcal{C}(\mathfrak{f})) = \chi \circ i(\mathfrak{g}) \circ N_{\mathfrak{f},\mathfrak{g}}\mathcal{C}(\mathfrak{f}) \subset \chi \circ i(\mathfrak{g})(\mathcal{C}(\mathfrak{g}))$, proving half of our claim. To see that the quotient is pseudo-null observe that it is annihilated by $[K(\mathfrak{f}\mathfrak{p}^\infty) : K(\mathfrak{g}\mathfrak{p}^\infty)] = m$ (since $C(\mathfrak{g}) \subset C(\mathfrak{f})$), and also by $\prod (1-\chi(\mathfrak{l})^{-1}\sigma_{\mathfrak{l}}|_{\Gamma'}^{-1})$, where the product is over all $\mathfrak{l}$ dividing $\mathfrak{f}$ but not $\mathfrak{g}$ (because of (6)). These two elements of $\mathbf{D}'[[\Gamma']]$ are relatively prime, and any noetherian module annihilated by two relatively prime elements is pseudo-null. So the claim is established. Now apply corollary 1.5. If $\mathfrak{g} \neq (1)$ we are done, because $\Lambda/\Lambda_0$ and $\Lambda/\Lambda_1$ are pseudo-null. If $\mathfrak{g} = (1)$ but $\chi \neq 1$ then $\chi(\Lambda_{00})$ contains $\chi(\tau) - 1$ for some $\tau \in H'$, $\chi(\tau) \neq 1$, and also $\gamma_0 - 1$. These two are again relatively prime, hence $\chi(\Lambda)/\chi(\Lambda_{00})$ is pseudo-null. Incidently we have shown that $\mu(1;\chi) \in \mathbf{D}'[[\Gamma']]$. Finally if $\chi = 1$, $\chi(\Lambda_{00})$ is the augmentation ideal of $\mathbf{D}'[[\Gamma']]$, which is generated by $\gamma_0 - 1$.

**1.11 Corollary.** *The main conjecture is equivalent to the statement that* $\mu(\mathfrak{g};\chi)$, *if* $\chi \neq 1$, *or* $(\gamma_0 - 1)\mu(1;1)$, *if* $\chi = 1$, *is a characteristic power series for* $\mathcal{X}(\mathfrak{f})_\chi$. *Here* $\mathfrak{g}$ *is the prime-to-*$\mathfrak{p}$ *part of* $\mathfrak{f}_\chi$.

While (16) makes the connection with the fundamental exact sequence (13) evident, this last formulation is more practical, because $\mu(\mathfrak{g};\chi)$ is nothing but the $p$-adic $L$ function of $\chi^{-1}$, restricted to characters of $Gal(K_\infty/K)$. In the notation of II.4.16, for any $p$-adic character $\rho$ of $\Gamma'$ (a character of "the second kind")

(18) $$L_{p,\mathfrak{g}}(\chi^{-1}\rho^{-1}) = \int_{\mathcal{G}(\mathfrak{g})} \chi\rho \, d\mu(\mathfrak{g}) = \int_{\Gamma'} \rho \, d\mu(\mathfrak{g};\chi).$$

As mentioned above, we shall see in the next section that $\mathcal{X}(\mathfrak{f})$ is free of finite rank as a $\mathbf{Z}_p$-module, and on the other hand $\mu(\mathfrak{g};\chi)$ (or $(\gamma_0-1)\mu(1;1)$ if $\chi = 1$) is not divisible by the uniformizer of $\mathbf{D}'$. From the general theory of Iwasawa modules, the main conjecture then reduces to the following statement:

> The eigencharacters of $\mathcal{G}(\mathfrak{f})$ in its action on the finite dimensional vector space $\mathcal{X}(\mathfrak{f}) \otimes \mathbf{Q}$ are precisely those $\varepsilon = \chi\rho$ for which (18) vanishes. I.e. they are the *reciprocals* of the zeroes of the (primitive) $p$-adic $L$ functions of conductors dividing $\mathfrak{f}$.

Compare with the discussion at the beginning of this chapter!

In one direction the main conjecture says that if we know a zero of the $p$-adic $L$ function, we should be able to construct an abelian $p$-extension of $F_\infty$ on which $H' \times \Gamma'$ acts in a prescribed way, and which is unramified outside $\mathfrak{p}$. How do we construct such an extension? A natural approach is through *Kummer theory* on elliptic curves with complex multiplication by $K$. The main conjecture is thereby intimately related to the arithmetic of such curves, a topic that will make up the next chapter.

**1.12** THE TRIVIAL CHARACTER: With scant information as we have about $\mathcal{X}(\mathfrak{f})$, any partial result is welcome. We shall show that at least at the trivial character ($\chi = \rho = 1$ in (18)) the main conjecture holds.

**Proposition.** *(i) $\mu(1)$ has a true pole at the trivial character.*

*(ii) $\rho = 1$ is not a zero of char. $\mathcal{X}(1)_1$.*

PROOF: (i) This is a restatement of corollary II.5.3. (ii) If $\rho = 1$ is a zero of $\mathcal{X}(1)_1$, there is a $\mathbf{Z}_p$ extension of $F_\infty = K(\mathfrak{p}^\infty)$ on which $Gal(F_\infty/K)$ acts trivially, and which is unramified outside $\mathfrak{p}$. Therefore there exists already a $\mathbf{Z}_p$ extension $N$ of this type over $K_\infty$. But such an $N$ would be abelian over $K$, because it is a trivial extension of a pro-cyclic group by an abelian one, and this is impossible.

Similarly, one can use the fact that Leopoldt's conjecture is known for abelian extensions of $K$ [Br] to prove the following.

**Proposition.** *Let $\varepsilon$ be a character of finite order of $K$, $\varepsilon \neq 1$, and $\mathfrak{f}$ the prime-to-$\mathfrak{p}$ part of $\mathfrak{f}_\varepsilon$. Then*

*(i) $\int_{\mathcal{G}(\mathfrak{f})} \varepsilon \, d\mu(\mathfrak{f}) \neq 0$.*

*(ii) $\varepsilon$ does not occur in the representation of $\mathcal{G}(\mathfrak{f})$ on $\mathcal{X}(\mathfrak{f})$.*

**1.13** The following result of Greenberg will be used later to show that the $p$-torsion of $\mathcal{X}(\mathfrak{f})$ is trivial.

**Theorem** ([GRE1], END OF §4). *The $\mathbf{Z}_p[[\mathcal{G}]]$-module $\mathcal{X}(\mathfrak{f})$ does not have any non-trivial finite submodules. (Finite means of finite cardinality).*

**1.14** THE "TWO-VARIABLE" MAIN CONJECTURE: When $\mathfrak{f}$ is replaced by $\mathfrak{f}\overline{\mathfrak{p}}^{\infty}$ one can write

$$(19) \qquad \mathcal{G}(\mathfrak{f}\overline{\mathfrak{p}}^{\infty}) \simeq H' \times \Gamma_1 \times \Gamma_2$$

where $\Gamma_1$ (resp. $\Gamma_2$) is the Galois group of the unique $\mathbf{Z}_p$ extension of $K$ unramified outside $\mathfrak{p}$ (resp. $\overline{\mathfrak{p}}$). Taking the inverse limit of the measures $\mu(\mathfrak{f}\overline{\mathfrak{p}}^{n})$ we obtain a measure $\nu(\mathfrak{f}) = \mu(\mathfrak{f}\overline{\mathfrak{p}}^{\infty})$ (see II.4.14) on $\mathcal{G}(\mathfrak{f}\overline{\mathfrak{p}}^{\infty})$, and for any $\chi \in \hat{H}'$

$$(20) \qquad \chi(\nu(\mathfrak{f})) = \nu(\mathfrak{f};\chi) \in \mathbf{D}'[[\Gamma_1 \times \Gamma_2]].$$

The structure theory for $\Gamma$-modules generalizes easily to $\Gamma_1 \times \Gamma_2$-modules ([Cu] §1 and IV.3.7) and we may make the conjecture that for any $\chi \in \hat{H}'$

$$(21) \qquad \nu(\mathfrak{f};\chi) = char.\, \mathcal{X}(\mathfrak{f}\overline{\mathfrak{p}}^{\infty})_{\chi},$$

provided $\mathfrak{f}$ is the prime-to-$p$ part of $\mathfrak{f}_{\chi}$.

## 2. THE IWASAWA INVARIANTS

**2.1** We retain the notation and assumptions of the previous section. In particular we fix a decomposition $\mathcal{G}(\mathfrak{f}) \simeq H' \times \Gamma'$ as in 1.8. Recall that any $h \in \mathbf{D}'[[\Gamma']]$ can be written uniquely in a Weierstrass form

$$(1) \qquad h = \pi^m h_w u$$

where $\pi$ is a uniformizer of $\mathbf{D}'$, $m$ a non-negative integer, $h_w$ a monic polynomial, all of whose coefficients except for the leading one are divisible by $\pi$, and $u$ a unit. Let $e$ be the absolute ramification index of $\mathbf{D}'$. The invariants

$$(2) \qquad \mu - inv(h) = \frac{m}{e}, \qquad \lambda - inv(h) = deg\, h_w$$

are called its *Iwasawa invariants*. If $M$ is a noetherian torsion $\mathbf{D}'[[\Gamma']]$-module, we refer to the invariants of $char.(M)$ as the invariants of $M$. If $M = M_0 \hat{\otimes} \mathbf{D}'$ is obtained from a $\mathbf{Z}_p[[\Gamma']]$-module by extension of scalars, its invariants coincide with those of $M_0$ (which we leave to the reader to define).

Let $f_\chi = char.\ \mathcal{X}(\mathfrak{f})_\chi$, $g_\chi = char.\ (\mathcal{U}(\mathfrak{f})/\mathcal{C}(\mathfrak{f}))_\chi$ be the two power series figuring in the main conjecture (1.9), for $\chi \in \hat{H}'$. We shall prove the following corollary of 1.9:

**Theorem.** *The total $\mu-$ and $\lambda$-invariants of $\mathcal{X}(\mathfrak{f})$ and $\mathcal{U}(\mathfrak{f})/\mathcal{C}(\mathfrak{f})$ are equal:*

$$\begin{aligned} \sum_\chi \mu - inv(f_\chi) &= \sum_\chi \mu - inv(g_\chi) \\ \sum_\chi \lambda - inv(f_\chi) &= \sum_\chi \lambda - inv(g_\chi). \end{aligned} \tag{3}$$

In fact, the $\mu$-invariants vanish (2.12), but this is proven for the $g_\chi$, and (3) is needed to get it for the $f_\chi$.

Apart from giving support for conjecture 1.9, (3) reduces its proof to showing $g_\chi | f_\chi$ for all $\chi$ (or the other way around). In the cyclotomic theory, the same phenomenon enabled Mazur and Wiles to deduce the cyclotomic main conjecture from a "$g_\chi | f_\chi$"-type result.

The proof of (3) goes as follows. Let $f$ be a characteristic power series for $\mathcal{X}(\mathfrak{f})$, considered as a $\Gamma'$-module, and $g = \prod g_\chi$. We shall compute the invariants of $f$ and $g$ separately, using class field theory in the case of the former, and the analytic class number formula for the latter, and conclude that they are equal. Quoting the important result of Gillard [Gi2] on the vanishing of the $\mu$-invariant of $g$, we deduce that the $p$-torsion of $\mathcal{X}(\mathfrak{f})$ is finite, hence trivial by 1.13. But then the $\mu$-invariant of each $\mathcal{X}(\mathfrak{f})_\chi$ vanishes too, and $f = \prod f_\chi$. This gives the identities in (3).

**2.2** For any group $G$, and any $G$-module $X$, let $X_G$ be the module of $G$-coinvariants, i.e. the largest quotient of $X$ on which $G$ acts trivially. With $\Gamma_n = \Gamma^{p^n} = \Gamma'^{p^{t+n}} = Gal(F_\infty/F_{n+1})$ $(n \geq 0)$,

$$Gal(M_{n+1}/F_\infty) = \mathcal{X}(\mathfrak{f})_{\Gamma_n}, \tag{4}$$

because $M_n$ is the largest abelian extension of $F_n$ inside $M_\infty$. To find the $\lambda$ and $\mu$ invariants of $\mathcal{X}(\mathfrak{f})$ we shall compute in 2.7 (11) below, following Coates-Wiles [C-W3], the order of (4). Then we use the following fundamental lemma of Iwasawa ([Wa] 13.3):

**Lemma.** *Let $X$ be a noetherian torsion $\mathbf{Z}_p[[\Gamma']]$-module, and $f$ its characteristic power series. Assume that $X_{\Gamma_n}$ $(\Gamma_n = \Gamma'_{t+n})$ is finite for all $n$. Then there exists a constant $\nu$ such that for large enough $n$*

$$(5) \qquad length(X_{\Gamma_n}) = \mu - inv(f) \cdot p^{t+n} + \lambda - inv(f) \cdot n + \nu.$$

(The invariants are as $\Gamma'$-modules. When we replace $\Gamma'$ by $\Gamma = \Gamma'_t$, the $\lambda$-invariant is unchanged and the $\mu$-invariant is multiplied by $p^t$. The length, in our case, is simply $ord_p \, \sharp(X_{\Gamma_n})$, because we have not yet extended scalars to $\mathbf{D}'$).

**2.3** $p$-ADIC REGULATORS: For the moment, let $F$ be any abelian extension of $K$ of degree $d$. Let $\sigma_1, \ldots, \sigma_d$ be those embeddings of $F$ in $\mathbf{C}_p$ that induce $\mathfrak{p}$ on $K$. Let $E$ be a subgroup of finite index in the full group of units of $F$, and choose generators $e_1, \ldots, e_{d-1}$ for $E/tor(E)$. Let log denote the $p$-adic logarithm.

DEFINITION. *The $\mathfrak{p}$-adic regulator of $E$ is*

$$(6) \qquad R_\mathfrak{p}(E) = det(\log \sigma_i(e_j))_{1 \le i,j \le d-1}.$$

It is well defined up to sign, because $\log N_{F/K}(e) = 0$ for $e$ in $E$. If we let $e_d = 1 + p$ we get (add the first $d-1$ rows to the last one):

$$(6') \qquad R_\mathfrak{p}(E) = (d \log \varepsilon_d)^{-1} \cdot det(\log \sigma_i(e_j))_{1 \le i,j \le d}.$$

Since we assumed that $F/K$ is abelian, Leopoldt's conjecture holds, namely,

**Theorem** (BAKER-BRUMER [BR]). $R_\mathfrak{p}(E) \neq 0$.

We shall denote by $disc_\mathfrak{p}(F/K)$ the $\mathfrak{p}$-part of the relative discriminant. It is an ideal in $K$, which is a certain power of $\mathfrak{p}$.

**2.4** For each prime $\mathfrak{P}$ of $F$ lying above $\mathfrak{p}$ let $w_{\mathfrak{P}}$ be the number of $p$-power roots of unity in the local field $F_{\mathfrak{P}}$. Let $\Phi = F \otimes K_{\mathfrak{p}} = \prod F_{\mathfrak{P}}$, and let $U$ be the group of principal units in $\Phi$ (compare II.4.1). The $p$-adic logarithm gives a homomorphism $\log : U \to \Phi$ whose kernel has order $\prod w_{\mathfrak{P}}$ (the $w_{\mathfrak{P}}$, for $\mathfrak{P}|\mathfrak{p}$, are actually equal). The image $\log(U)$ is an open subgroup of $\Phi$.

Let $E$ be a subgroup of finite index in $\mathcal{O}_F^x$ as above, and $D = E \cdot <1+p>$. Let $\overline{D}$ be its closure in the units of $\Phi$, and $<\overline{D}>$ the projection of $\overline{D}$ to $U$.

**Lemma**. *The index of* $\log(<\overline{D}>)$ *in* $\log(U)$ *is finite, and given by*

$$(7) \qquad ord_p[\log(U) : \log(<\overline{D}>)] = ord_{\mathfrak{p}}\left(\frac{dpR_{\mathfrak{p}}(E)}{\sqrt{disc_{\mathfrak{p}}(F/K)}} \cdot \prod_{\mathfrak{P}|\mathfrak{p}} (w_{\mathfrak{P}}\mathrm{N}\mathfrak{P})^{-1}\right).$$

PROOF: See [C-W3], lemma 8.

**2.5 Corollary**. *Let $w(E)$ be the number of roots of unity in $E$. Then*

$$(8) \qquad ord_p[U : <\overline{D}>] = ord_{\mathfrak{p}}\left(\frac{dpR_{\mathfrak{p}}(E)}{w(E)\sqrt{disc_{\mathfrak{p}}(F/K)}} \cdot \prod_{\mathfrak{P}|\mathfrak{p}} \mathrm{N}\mathfrak{P}^{-1}\right).$$

PROOF: An immediate consequence of (7) ([C-W3] lemma 9).

**2.6** We return to the situation considered in theorem 2.1. With the notation of 1.7, the Artin symbol induces an isomorphism

$$(9) \qquad U_n/ <\overline{E}_n> \cong Gal(M_n/L_n).$$

Recall that $L_n$ (resp. $M_n$) is the maximal abelian unramified (resp. unramified outside $\mathfrak{p}$) $p$-extension of $F_n$. Clearly $F_\infty \subset M_n$ and $L_n \cap F_\infty = F_n$.

Let $Y_n \subset U_n$ be the subgroup for which

$$(10) \qquad Y_n/ <\overline{E}_n> \cong Gal(M_n/L_nF_\infty).$$

**Lemma**. $Y_n = Ker\ N_{F_n/K}|U_n$.

PROOF: The norm is of course from $\Phi_n = F_n \otimes K_\mathfrak{p}$ to $K_\mathfrak{p}$. If $u \in U_n$ and $(u, F_\infty/F_n) = 1$, then $(N_{F_n/K}u, F_\infty/K) = 1$, so the idele $N_{F_n/K}u$, which is 1 outside $\mathfrak{p}$, is necessarily 1. This argument can be reversed to prove the converse.

**2.7 Proposition** ([C-W3] THEOREM 11). *For any field $F$ containing $K$ let $h(F)$ be its class number and $R_\mathfrak{p}(F)$ the $\mathfrak{p}$-adic regulator of $\mathcal{O}_F^x$. Let $w_F$ be the number of roots of unity in $F$. Then in the situation considered above $M_n/F_\infty$ is a finite extension ($n \geq 1$) and*

$$(11)\quad ord_p[M_n : F_\infty] = ord_\mathfrak{p}\left\{\frac{p^n h(F_n) R_\mathfrak{p}(F_n)}{w_{F_n}\cdot\sqrt{disc_\mathfrak{p}(F_n/K)}}\cdot\prod_{\mathfrak{P}|\mathfrak{p}}(1 - \mathrm{N}\mathfrak{P}^{-1})\right\}.$$

*(the product is over the primes $\mathfrak{P}$ of $F_n$).*

PROOF: Let $p^\delta \| [F : K]$, so $p^{n+\delta-1} \| [F_n : K]$, for $n \geq 1$. Let $D_n = E_n \cdot <1+p>$, and note that since $N_{F_n/K}(E_n) = 1$, $N_{F_n/K}(<\overline{D}_n>) = 1 + p^{n+\delta}\mathcal{O}_\mathfrak{p}$. From local class field theory, $N_{F_n/K}(U_n) = 1 + \mathfrak{p}^n \mathcal{O}_\mathfrak{p}$. Consider the diagram

$$\begin{array}{ccccccccc} 0 & \to & <\overline{E}_n> & \to & <\overline{D}_n> & \to & 1+p^{n+\delta}\mathcal{O}_\mathfrak{p} & \to & 0 \\ & & \downarrow & & \downarrow & & \downarrow & & \\ 0 & \to & Y_n & \to & U_n & \xrightarrow{N_{F_n/K}} & 1+p^n\mathcal{O}_\mathfrak{p} & \to & 0 \end{array}$$

with exact rows and injective columns. Since $d = [F_n : K]$ is exactly divisible by $p^{n+\delta-1}$, (11) is a consequence of (8), (10), and the fact that $[L_nF_\infty : F_\infty] = [L_n : F_n] =$ the $p$-part of $h(F_n)$.

**2.8 Corollary.** *Let $f$ be a characteristic power series for $\mathcal{X}(\mathfrak{f})$ as a $\Gamma'$-module. Then for $n >> 0$*

$$(12)\quad \begin{aligned} &\mu - inv(f)\cdot p^{t+n-1}(p-1) + \lambda - inv(f) = \\ &\qquad 1 + ord_\mathfrak{p}\left\{\frac{hR_\mathfrak{p}}{w\sqrt{disc_\mathfrak{p}}}(F_{n+1}) \Big/ \frac{hR_\mathfrak{p}}{w\sqrt{disc_\mathfrak{p}}}(F_n)\right\}. \end{aligned}$$

PROOF: By the previous proposition, the right hand side is equal to $ord_p([M_{n+1} : F_\infty]/[M_n : F_\infty])$. Now apply (4) and lemma 2.2 (5).

**2.9** We now begin the computation of the Iwasawa invariants of the $p$-adic $L$ functions $g_\chi$. Let $g = \prod g_\chi$, the product taken over all $\chi \in \hat{H}'$. We have the following useful result.

**Lemma.** *For any character of finite order $\rho$ of $\Gamma'$, let level $(\rho) = s$ if $\rho(\Gamma'^{p^s}) = 1$, but $\rho(\Gamma'^{p^{s-1}}) \neq 1$. Then for $n >> 0$*

(13)
$$\mu - inv(g) \cdot p^{t+n-1}(p-1) + \lambda - inv(g) = ord_p \left\{ \prod_{level(\rho)=t+n} \rho(g) \right\}.$$

*Here $ord_p$ is the valuation of $\mathbf{C}_p$ normalized by $ord_p(p) = 1$.*

PROOF: We may enlarge $\mathbf{D}'$ to assume that $g$ has all its zeroes there. This reduces the proof of (13) to the two special cases $g = \pi$ (a uniformizer of $\mathbf{D}'$) or $g = \gamma_0 - 1 - \alpha$ ($\gamma_0$ a topological generator of $\Gamma'$, $|\alpha| < 1$). Both are easy exercises.

**2.10 Proposition.** *For any ramified character $\varepsilon$ of $Gal(F_\infty/K)$ let $\mathfrak{g} = \mathfrak{f}_\varepsilon$, $(g) = \mathfrak{g} \cap \mathbf{Z}$, and*

(14)
$$S_p(\varepsilon) = -\frac{1}{12gw_\mathfrak{g}} \cdot \sum_{C\in Cl(\mathfrak{g})} \varepsilon^{-1}(C) \log \varphi_\mathfrak{g}(C)$$

*(compare II.5.2 (2)). Define $G(\varepsilon)$ as in II.4.11 (30). Let $A_n$ be the collection of all $\varepsilon$ for which $\mathfrak{p}^n \| \mathfrak{f}_\varepsilon$. Then, for large $n$,*

(15)
$$ord_p\left(\prod_{\varepsilon\in A_n} G(\varepsilon)S_p(\varepsilon)\right) = ord_p\left[\frac{hR_\mathfrak{p}}{w\sqrt{disc_\mathfrak{p}}}(F_n) \Big/ \frac{hR_\mathfrak{p}}{w\sqrt{disc_\mathfrak{p}}}(F_{n-1})\right].$$

The proof of the proposition will be given in 2.11. Before embarking upon it, we conclude the proof of the equality of the invariants of $f$ and $g$.

Every $\varepsilon \in A_{n+1}$ can be written as $\varepsilon = \chi\rho$, where $\chi$ is a character of $H'$, and $\rho$ of $\Gamma'/\Gamma'_{t+n}$ but not of $\Gamma'/\Gamma'_{t+n-1}$. Theorem II.5.2 shows that

$$\rho(g_\chi) = \int_{\mathcal{G}(\mathfrak{g}_0)} \chi\rho \, d\mu(\mathfrak{g}_0) = G(\varepsilon)S_p(\varepsilon), \qquad \text{if } \chi \neq 1$$

$(\mathfrak{g} = \mathfrak{f}_\varepsilon = \mathfrak{g}_0\mathfrak{p}^n)$. If $\chi = 1$, we similarly get

$$\rho(g_\chi) = (\rho(\gamma_0) - 1)G(\varepsilon)S_p(\varepsilon)$$

where $\gamma_0$ is a topological generator of $\Gamma'$. When we take the product over all $\varepsilon \in A_{n+1}$, we take all $\chi \in \hat{H}'$, and all $\rho \in \hat{\Gamma}'$ of level $t + n$. Therefore (15) yields

$$(16) \qquad \prod_{level(\rho)=t+n} \rho(g) \sim p \cdot \frac{hR_\mathfrak{p}}{w\sqrt{disc_\mathfrak{p}}}(F_{n+1}) \Big/ \frac{hR_\mathfrak{p}}{w\sqrt{disc_\mathfrak{p}}}(F_n)$$

where $\sim$ means "up to a $p$-adic unit". Here we used the fact that $\prod(\rho(\gamma_0) - 1) = \prod(\varsigma - 1) \sim p$ if $\varsigma$ runs over all primitive roots of unity of order $p^{n+t}$.

Comparing (16) with (13), and then with (12), finally shows that $f$ and $g$ have the same invariants.

**2.11** To prove (15), observe first that

$$(17) \qquad \prod_{\varepsilon\in A_n} G(\varepsilon) \sim \{disc_\mathfrak{p}(F_{n-1}/K) \,/\, disc_\mathfrak{p}(F_n/K)\}^{1/2}.$$

This is nothing but the "conductor-discriminant formula" localized at the prime $\mathfrak{p}$. It remains to show, therefore, that

$$(18) \qquad \prod_{\varepsilon\in A_n} S_p(\varepsilon) \sim \frac{hR_\mathfrak{p}}{w}(F_n) \Big/ \frac{hR_\mathfrak{p}}{w}(F_{n-1}).$$

To this end, introduce, for a ramified $\varepsilon$, the sum

$$(19) \qquad S_\infty(\varepsilon) = -\frac{1}{12gw_\mathfrak{g}} \cdot \sum_{C\in Cl(\mathfrak{g})} \varepsilon^{-1}(C) \log|\varphi_\mathfrak{g}(C)|^2$$

(the logarithm here is the ordinary logarithm, and $\mathfrak{g}$ and $g$ are as in (14)). If $\varepsilon$ is unramified, but non-trivial, let

$$(14') \qquad S_p(\varepsilon) = \frac{1}{12h_K w_K} \cdot \sum_{C\in Cl(1)} \varepsilon^{-1}(C) \log \delta(C)$$

$$(19') \qquad S_\infty(\varepsilon) = \frac{1}{12h_K w_K} \cdot \sum_{C\in Cl(1)} \varepsilon^{-1}(C) \log |\delta(C)|^2$$

where $\delta(C)$ is Siegel's unit (II.2.2).

Let $H_n = Gal(F_n/K)$, $I(H_n)$ = the augmentation ideal in $\mathbf{Z}[H_n]$, and choose an embedding $\eta$ of $I(H_n)$ into the units of $F_n$, as Galois modules. The image of $\eta$ is a subgroup $E'_n$ of finite index in $E_n$. If we put

$$\Sigma_p(\varepsilon) = \sum_{\sigma\in H_n} \varepsilon^{-1}(\sigma) \log \eta(\sigma) \qquad (p\text{-adic log})$$

$$\Sigma_\infty(\varepsilon) = \sum_{\sigma\in H_n} \varepsilon^{-1}(\sigma) \log |\eta(\sigma)|^2 \qquad (\text{ordinary log})$$

($\varepsilon \neq 1$) then there is a non-zero algebraic number $r_\varepsilon$ such that

$$(20) \qquad S_p(\varepsilon) = r_\varepsilon \cdot \Sigma_p(\varepsilon), \qquad S_\infty(\varepsilon) = r_\varepsilon \cdot \Sigma_\infty(\varepsilon).$$

Now, the *analytic class number formula* together with Kronecker's theorem (as formulated by Siegel) II.5.1, gives

$$(21) \qquad \prod_{\varepsilon\neq 1, \varepsilon\in \hat{H}_n} S_\infty(\varepsilon) = \alpha \cdot \frac{hR_\infty}{w}(F_n)$$

where $\alpha$ is a constant independent of $n$. Here $R_\infty$ is the usual (complex) regulator, $h$ the class number and $w$ the number of roots of unity. In the following computation $\varepsilon$ ranges over all non-trivial characters of $Gal(F_n/K)$. Use has been made of the fact that $R_\mathfrak{p}(E'_n)/R_\mathfrak{p}(E_n) = R_\infty(E'_n)/R_\infty(E_n) = [E_n : E'_n\mu_{F_n}]$. We find out

$$\prod S_\mathfrak{p}(\varepsilon) = \prod \Sigma_p(\varepsilon) \cdot \left(\frac{S_\infty(\varepsilon)}{\Sigma_\infty(\varepsilon)}\right) \qquad (\text{from (20)})$$

$$= R_\mathfrak{p}(E'_n)\left(\frac{\prod S_\infty(\varepsilon)}{R_\infty(E'_n)}\right) \qquad (\text{Frobenius determinant})$$

$$= R_\mathfrak{p}(E_n) \cdot \left(\frac{\prod S_\infty(\varepsilon)}{R_\infty(E_n)}\right)$$

$$= \alpha \cdot \frac{hR_\mathfrak{p}}{w}(F_n) \qquad (\text{from (21)}).$$

Dividing these expressions for $n - 1$ and for $n$ we arrive at (18), as desired.

**2.12** In [Gi2] R. Gillard uses ideas of W. Sinnott to prove the following. (L. Schneps proved the same result independently too.)

**Theorem** ([GI2] 2.9). *Fix a decomposition $\mathcal{G}(\mathfrak{f}) = H' \times \Gamma'$ as above, and let $g_\chi$ be defined by the right hand side of 1.10 (17), as before. Then $\mu - inv(g_\chi) = 0$ $(\forall \chi \in \hat{H}')$.*

As explained at the beginning, this concludes the proof of theorem 2.1. It also has the following

**Corollary.** *$\mu - inv(f_\chi) = 0$ for every $\chi \in \hat{H}'$. The $\mathcal{G}(\mathfrak{f})$-module $\mathcal{X}(\mathfrak{f})$ is a free $\mathbf{Z}_p$-module of finite rank.*

Gillard's theorem is another example, so common in number theory, where a purely algebraic statement (as the last corollary) is proved by analytical means.

## 3. Further Topics

In this short section we make a few more comments concerning the main conjecture, without proofs.

**3.1** Relations in ideal class groups: Building upon ideas of F. Thaine, K. Rubin proved recently a remarkable theorem in the direction of the main conjecture. Among other things he showed that in the fundamental exact sequence 1.7(13), the module $\mathcal{E}/\mathcal{C}$ "governs" $\mathcal{W}$. So far the results depend on various simplifying assumptions, which we proceed to describe, but these seem to be removable.

Assume that $K$ has class number 1, and that $E$ is an elliptic curve defined over $K$ with complex multiplication by $\mathcal{O}_K$, and conductor $\mathfrak{f}$. Let $p$ be a split prime of good reduction, and assume $p > 2$. Consider the fields $F_n = K(E[\mathfrak{p}^n])$. Our notation is different from that employed previously, but note (II.1.6) that $F_n \subset K(E[\mathfrak{f}\mathfrak{p}^n])) = K(\mathfrak{f}\mathfrak{p}^n)$. Write, as usual, $Gal(F_\infty/K) = \Gamma \times \Delta$, and let

$\chi \in \hat{\Delta}$. The main conjecture for $\chi$ asserts that $(\mathcal{U}(\mathfrak{f})/\mathcal{C}(\mathfrak{f}))^\chi$ and $\mathcal{X}(\mathfrak{f})^\chi$ have the same characteristic ideal as $\mathbf{Z}_p[[\Gamma]]$-modules. By 1.7(13), this is equivalent with

$$(1) \qquad char.(\mathcal{E}(\mathfrak{f})/\mathcal{C}(\mathfrak{f}))^\chi = char.\ \mathcal{W}(\mathfrak{f})^\chi.$$

**Theorem** ([RU3]). *Let* $h_\chi = char.(\mathcal{E}(\mathfrak{f})/\mathcal{C}(\mathfrak{f}))^\chi$. *Then* $h_\chi \cdot \mathcal{W}(\mathfrak{f})^\chi$ *is pseudo-null (i.e. finite).*

This comes close to (1). Since we know that the total $\lambda$-invariants (summed over all $\chi$) of the two modules in (1) are the same, and their $\mu$-invariants vanish, (1) would follow from the theorem if all the zeroes of $char.\ \mathcal{W}(\mathfrak{f})^\chi$ are simple, for all $\chi$. More generally, it would be enough to know that $\mathcal{W}(\mathfrak{f})^\chi$ is a cyclic $\mathbf{Z}_p[[\Gamma]]$-module, for all $\chi$.

**3.2** THE FUNCTIONAL EQUATION: We have seen in II.6 that the $p$-adic $L$ function satisfies a functional equation with respect to $\varepsilon \mapsto \check{\varepsilon}$, and in particular $L_p(\varepsilon) = 0 \Leftrightarrow L_p(\check{\varepsilon}) = 0$, where the modulus of the $p$-adic $L$ function consists of the ramified primes for $\varepsilon$, as well as $\mathfrak{p}$ and $\bar{\mathfrak{p}}$. In view of the discussion in 1.11 we would like to say that $\varepsilon$ occurs in the representation of $\mathcal{G}(\mathfrak{f}\bar{\mathfrak{p}}^\infty)$ on $\mathcal{X}(\mathfrak{f}\bar{\mathfrak{p}}^\infty)$ precisely when $\check{\varepsilon}$ occurs in the representation of $\mathcal{G}(\bar{\mathfrak{f}}\bar{\mathfrak{p}}^\infty)$ on $\mathcal{X}(\bar{\mathfrak{f}}\bar{\mathfrak{p}}^\infty)$. This is indeed true, but to prove it one has to await a different description of the module $\mathcal{X}$ (cf. IV.1.5). Even then the "algebraic functional equation" is a deep fact. See [PR1] V, §1, or [Maz2] §7.

**3.3** A KUMMER CRITERION FOR QUADRATIC IMAGINARY FIELDS: In analogy with the well known notion of a regular prime (in $\mathbf{Q}$), one may define a split prime $\mathfrak{p}$ of $K$ to be *regular* for $F = K(\mathfrak{f})$ ($(\mathfrak{f}, \mathfrak{p}) = 1$ as before) if the module $\mathcal{X}(\mathfrak{f})$ is trivial. Since by Nakayama's lemma $\mathcal{X}(\mathfrak{f}) = (0)$ if and only if $\mathcal{X}(\mathfrak{f})_\Gamma = (0)$ ($\Gamma = Gal(F_\infty/F_1)$), and since $\mathcal{X}(\mathfrak{f})_\Gamma \cong Gal(M_1/F_\infty)$ (see 2.2(4)), $\mathfrak{p}$ is regular for $F$ if and only if the only cyclic extension of degree $p$ of $K(\mathfrak{f}\mathfrak{p})$, unramified outside $\mathfrak{p}$, is $K(\mathfrak{f}\mathfrak{p}^2)$. Compare with the cyclotomic theory. There

$p$ is regular (for $\mathbf{Q}$) if the only cyclic extension of degree $p$ of $\mathbf{Q}(p) = \mathbf{Q}(\cos \frac{2\pi}{p})$, unramified outside $p$, is $\mathbf{Q}(p^2)$.

Theorem 2.1, in conjunction with 1.13, shows that a necessary and sufficient condition for $\mathfrak{p}$ to be regular for $K(\mathfrak{f})$, is that the power series $g_\chi$, $\chi \in \hat{H}'$, are units. (See 1.8 for notation, and recall that $g_\chi$ is defined by 1.9(17)). It is now easy to obtain a criterion for regularity in terms of divisibility of special values of $L$ functions by $p$. When $K$ has class number 1, this was done by Coates and Wiles (if $\mathfrak{f} = 1$, [C-W3]), and by Yager, in a more general context ([Ya3], theorem 3). Note that there is no reason to stick to $F$ which is a ray class field. The definition of "regular for $F$", and the corresponding criterion, carry over to any abelian extension of $K$.

# CHAPTER IV

# APPLICATIONS TO THE ARITHMETIC OF ELLIPTIC CURVES WITH COMPLEX MULTIPLICATIONS

This chapter deals with the conjecture of Birch and Swinnerton-Dyer. The conjecture relates arithmetical invariants associated to the group of $F$-rational points on an elliptic curve $E$ (defined over a number field $F$), to analytic invariants derived from its Hasse-Weil zeta function $L(E/F, s)$. It sprung from extensive computations done in the early sixties on the curves $y^2 = x^3 - Dx$ [B-SD] and $x^3 + y^3 = D$ [St]. Both families of curves have complex multiplication, by $\mathbf{Q}(i)$ and $\mathbf{Q}(\sqrt{-3})$ respectively. Tate has studied far-reaching generalizations in [Ta4], but until 1976 very little was known.

The last decade has witnessed a few remarkable breakthroughs. Coates and Wiles [C-W1] and R. Greenberg [Gre2] used $p$-adic techniques and Iwasawa theory to obtain results about elliptic curves with complex multiplication. Gross and Zagier [G-Z] used special points on modular curves to obtain a theorem pertaining to any elliptic curve defined over $\mathbf{Q}$, and uniformized by a modular curve. These include the curves with potential complex multiplication which are defined over $\mathbf{Q}$. Recently, K. Rubin combined new $p$-adic results with the three works cited above, together with important work of B. Perrin-Riou on $p$-adic heights [PR2], to obtain stronger theorems in the CM case. He also gave the first examples of finite Tate-Shafarevitch groups, thereby verifying the full conjecture for the first time.

In this chapter we shall give a complete account on the theorems of Coates-Wiles and Greenberg. These theorems were originally proven for curves defined over their field of complex multiplication. Here we give generalizations to quadratic imaginary fields of higher class number, and curves satisfying II.1.4(12). The generalizations are due to N. Arthuand [Art] and K. Rubin [Ru1] for the Coates-Wiles theorem, and to the author for Greenberg's theorem, but they are mainly variations on the original themes.

The work of Gross and Zagier uses an entirely different circle of ideas. We refer the interested reader to their paper.

The latest developments mentioned above, although a natural continuation of this chapter, are not treated here. As this book goes to press, the work of K. Rubin is in preparation, and will appear shortly.

## 1. Descent and the Conjecture of Birch and Swinnerton-Dyer

In this section we describe the problem and the method of attack. Sections 1.1-1.3 are general, and are intended as background material. Starting in 1.4 we specialize to elliptic curves with complex multiplication. The key result is Coates' theorem relating the Selmer group to the Iwasawa module denoted by $\mathcal{X}$ in III.1. Our treatment of descent is far from complete, and we give the bare minimum required for §3. For a comprehensive study of descent and Iwasawa theory see [C] and [C-Go].

**1.1** Let $F$ be a number field, and $E$ an elliptic curve defined over $F$. The group $E(F)$ of $F$-rational points of $E$ is a finitely generated abelian group ([Sil] VIII, 6.7). It is called the *Mordell-Weil group*, and its rank is the first important invariant attached to $E/F$.

The *Hasse-Weil zeta function* of $E$ over $F$ is defined by the Euler product

$$L(E/F,s) = \prod_P L_P(s) \tag{1}$$

where $P$ runs over all the non-archimedean primes of $F$. If $P$ is a prime of good reduction (II.1.8) $L_P(s)$ is the reciprocal of the numerator of the zeta function of the reduced curve:

$$L_P(s) = \{(1 - \alpha \mathrm{N}P^{-s})(1 - \alpha' \mathrm{N}P^{-s})\}^{-1} \tag{2}$$

where $\alpha$ and $\alpha'$ are the two roots of the characteristic polynomial of the Frobenius automorphism (relative to $\mathcal{O}_F/P$). If $P$ is a prime of bad additive reduction we set $L_P(s) = 1$. If $P$ is a prime of bad multiplicative reduction, $L_P(s) = (1 \pm \mathbf{N}P^{-s})^{-1}$ where the $-$ sign is chosen if the tangents at the node of $\tilde{E}(mod\ P)$ are rational over $\mathcal{O}_F/P$, and the $+$ sign otherwise.

The product (1) converges absolutely in $Re(s) > \frac{3}{2}$. This is a consequence of the Riemann hypothesis ($|\alpha| = \sqrt{\mathbf{N}P}$ in (2)), proven by Hasse in his 1934 dissertation. It is conjectured that (1) admits analytic continuation to the whole complex plane, and satisfies a functional equation with respect to $s \mapsto 2 - s$ (see [Ta]).

If $E$ has complex multiplication by a quadratic imaginary field, then all the bad places are places of additive reduction, so in (1) we may simply ignore them. In that case it is also known that $L(E/F, s)$ admits analytic continuation and functional equation. Indeed, if $F \supset K$ and $\psi = \psi_{E/F}$ is the grossencharacter of $E/F$ (see II.1.3), then

$$(3) \qquad L(E/F, s) = L(\psi, s) \cdot L(\overline{\psi}, s),$$

so the zeta function is expressed as a product of two Hecke $L$-series. Note that if, in addition, $F/K$ is abelian, and $E$ satisfies (II.1.4(12)), so that $\psi = \varphi \circ N_{F/K}$ for some grossencharacter $\varphi$ of $K$, then

$$(4) \qquad L(\psi, s) = \prod_{\chi \in \widehat{Gal(F/K)}} L(\chi\varphi, s).$$

**1.2** In general, we have the following famous conjecture.

CONJECTURE OF BIRCH AND SWINNERTON-DYER (PART 1):

$$(5) \qquad rk\ E(F) = ord_{s=1}\ L(E/F, s).$$

Thus the order of the zero at $s = 1$ (where in general $L(E/F, s)$ is not known to exist) should yield the most interesting arithmetical invariant associated to $E$ over $F$, namely its rank.

If $E$ has complex multiplication by $\mathcal{O}_K$, and $F \supset K$, then $E(F)$ is an $\mathcal{O}_K$-module, and clearly (5) is equivalent to

$$(5') \qquad rk_{\mathcal{O}_K}\ E(F) = ord_{s=1}\ L(\psi, s).$$

**1.3** There is a second part to the conjecture of Birch and Swinnerton-Dyer, giving the leading coefficient of the Taylor expansion of $L(E/F, s)$ at $s = 1$ in terms of other arithmetical invariants associated to $E$ over $F$. See [Ta4] for details. Here we are solely concerned with (5). To study the rank of $E(F)$ one usually uses *descent*, a method that we shall now briefly sketch. See [Sil], chapter X, for a fuller account and examples.

Let $\alpha \in End(E/F)$. The *Kummer exact sequence*

$$(6) \qquad 0 \to E[\alpha] \to E(\overline{F}) \xrightarrow{\alpha} E(\overline{F}) \to 0$$

gives an exact sequence in Galois cohomology

$$(7) \qquad 0 \to E(F)/\alpha E(F) \to H^1(G_F, E[\alpha]) \to H^1(G_F, E(\overline{F}))[\alpha] \to 0.$$

Here $G_F = Gal(\overline{F}/F)$. The group $H^1(G_F, E(\overline{F}))$, called the *Weil-Chatelet group*, classifies principal homogeneous spaces for $E$ over $F$.

For each prime $P$ of $F$ choose once and for all a prime $\overline{P}$ of $\overline{F}$ lying above it. Let $G_P \subset G_F$ be the decomposition group of $\overline{P}/P$ and identify the algebraic closure of $F_P$ with the corresponding subfield of the completion of $\overline{F}$ at $\overline{P}$. In this discussion we have to include the archimedean primes, but if $F$ is totally imaginary they have no effect on what follows.

For each $P$ there is a sequence like (7) with $F_P$ replacing $F$. In particular we may consider the local Weil-Chatelet groups $H^1(G_P, E(\overline{F}_P))$. The *Tate-Shafarevitch group* $Ш(E/F)$ is by definition

$$(8) \qquad Ш(E/F) = Ker(H^1(G_F, E(\overline{F})) \xrightarrow{Res} \amalg_P \; H^1(G_P, E(\overline{F}_P))),$$

where the direct sum is over *all* $P$. It parametrizes locally trivial principal homogeneous spaces for $E$ over $F$.

The pre-image of $Ш(E/F)[\alpha]$ in $H^1(G_F, E[\alpha])$ is called the *$\alpha$-Selmer group*, $S_\alpha(E/F)$. We therefore obtain the *$\alpha$-descent exact sequence*

$$(9) \qquad 0 \to E(F)/\alpha E(F) \to S_\alpha(E/F) \to Ш(E/F)[\alpha] \to 0.$$

The Tate-Shafarevitch group is conjectured to be finite, and today is even known to be so in some cases, for example for many curves of the form $Y^2 = X^3 - DX$ [Ru3]. The advantage of (9) over (7) is that if we show $Ш(E/F)[\alpha] = 0$, we may compute the rank of $E(F)$ from $S_\alpha(E/F)$ (see Silverman's book and [Ru2] for examples). The full Weil-Chatelet group, on the other hand, is huge and uncontrollable.

**1.4** From now on assume that $F$ contains $K$, our quadratic imaginary field, and that $E$ has complex multiplication by $\mathcal{O}_K$. Comparing the exact sequences (9) for $\alpha^n$, $n \geq 1$, with a fixed $\alpha \in \mathcal{O}_K$, we get in the limit

$$(10)\quad 0 \to E(F) \otimes_{\mathcal{O}_K} \varinjlim \alpha^{-n}\mathcal{O}_K/\mathcal{O}_K \to S_{\alpha^\infty}(E/F) \to Ш(E/F)[\alpha^\infty] \to 0.$$

The group $S_{\alpha^\infty}(E/F)$ is a subgroup of $H^1(G_F, E[\alpha^\infty])$. The first map sends $u \otimes \alpha^{-n}\beta$ ($u \in E(F)$, $\beta \in \mathcal{O}_K$) to the cohomology class of the cocycle $\{\sigma \mapsto \sigma v - v\}$, where $v \in E(\overline{F})$ satisfies $\alpha^n(v) = \beta(u)$.

Before we state the next theorem let us make a general remark about Galois cohomology. Suppose $L$ is an algebraic extension of $F$, but not necessarily finite, and $R$ any $G_F$-module. Then $G_L$ is a closed subgroup of $G_F$ and

$$(11)\qquad H^1(G_L, R) = \varinjlim H^1(G_M, R),$$

where $M$ runs over all intermediate fields $F \subset M \subset L$, $[M : F] < \infty$, and the limit is taken with respect to restriction maps. This is clear if we keep in mind that all the cocycles intervening are continuous for the *discrete* topology on $R$. We agree to define $Ш(E/L)$ and $S_\alpha(E/L)$ as the inductive limits of $Ш(E/M)$ and $S_\alpha(E/M)$, $F \subset M \subset L$, $[M : F] < \infty$. (The subtle point here is that if $P$ is a place of $L$, the completion $L_P$ is not necessarily $\cup M_P$).

**1.5** Let $\mathfrak{p}$ be a split prime of $K$, unramified in $F$, such that $E$ has good (necessarily ordinary) reduction at every place of $F$ above $\mathfrak{p}$. Pick any $\alpha \in \mathcal{O}_K$ with $(\alpha) = \mathfrak{p}^h$. Then (10) becomes

$$(12)\qquad 0 \to E(F) \otimes_{\mathcal{O}_K} K_\mathfrak{p}/\mathcal{O}_\mathfrak{p} \to S(E/F)(\mathfrak{p}) \to Ш(E/F)(\mathfrak{p}) \to 0,$$

where $D(\mathfrak{p})$ denotes the $\mathfrak{p}$-primary component of the divisible group $D$. Let $F_n = F(E[\mathfrak{p}^n])$, $n \geq 0$.

**Theorem** (COATES). *There is a canonical isomorphism of Galois-modules*

$$S(E/F_\infty)(\mathfrak{p}) \simeq Hom(\mathcal{X}(F_\infty), E[\mathfrak{p}^\infty]),$$

*where $\mathcal{X}(F_\infty)$ is the Galois group of the maximal abelian p-extension of $F_\infty$ unramified outside $\mathfrak{p}$, over $F_\infty$.*

PROOF: $S(E/F_\infty)(\mathfrak{p})$ is by definition a subgroup of $H^1(G_{F_\infty}, E[\mathfrak{p}^\infty]) = Hom(\mathcal{H}, E[\mathfrak{p}^\infty])$ where $\mathcal{H}$ is the Galois group of the maximal abelian $p$-extension of $F_\infty$ over $F_\infty$. We therefore have to identify the homomorphisms that belong to the Selmer group with those which are unramified outside $\mathfrak{p}$.

First observe that over $F_1$, $E$ has good reduction everywhere. Indeed, $F_\infty$ is the compositum of $F_1$ with $K_\infty$, the unique $\mathbf{Z}_p$-extension of $K$ unramified outside $\mathfrak{p}$, as follows easily from the main theorem of complex multiplication. Since $F_\infty/F_1$ is unramified outside $\mathfrak{p}$, the criterion of Ogg-Néron-Shafarevitch (II.1.8) shows that $E$ has good reduction outside $\mathfrak{p}$ over $F_1$. By our assumption on $\mathfrak{p}$, $E/F_1$ has good reduction everywhere.

Suppose now that $P$ is a place of $F_\infty$ not above $\mathfrak{p}$, and $f \in S(E/F_\infty)(\mathfrak{p})$. Then there exists a point $v \in E(\overline{F}_P)$ such that for any $\sigma$ in the decomposition group $G_P \subset Gal(\overline{F}_\infty/F_\infty)$, $f(\sigma) = \sigma(v) - v \in E[\mathfrak{p}^\infty]$. The $\mathfrak{p}^\infty$-torsion points of $E$ map injectively under reduction *mod* $P$, because $P \nmid \mathfrak{p}$, and there is good reduction. If $I_P$ denotes the inertia group in $G_P$, and $\sigma \in I_P$, $\widetilde{f(\sigma)} = \tilde{\sigma}(\tilde{v}) - \tilde{v} = 0$, so $f(\sigma) = 0$ ($\tilde{v}$ is the image of $v$ under the reduction map). It follows that $f$ is unramified at $P$.

The converse is more difficult. Suppose $f$ is a homomorphism which is unramified outside $\mathfrak{p}$. Let $P$ be a prime of $F_\infty$ not above $\mathfrak{p}$. If we now denote by $G_P$ and $I_P$ its decomposition and inertia groups *in* $\mathcal{H}$, $G_P = I_P$, because $\mathcal{H}$ is a pro-$p$ extension, and the residue field of $P$, while not algebraically closed, has no $p$ extensions. ($P$ is inert in $F_\infty/F_m$ for some large $m$, and a finite field has a unique extension of degree $p^r$ for each $r$). Thus $f|G_P = 0$, so the local condition at $P$ for $f$ to be in the Selmer group, is met.

There remain the primes $P$ of $F_\infty$ dividing $\mathfrak{p}$. We shall show that for such $P$, $H^1(G_P, E(\overline{F}_P))(\mathfrak{p}) = 0$, so the local condition at $P$ is trivially satisfied, completing the proof that $f \in S(E/F_\infty)(\mathfrak{p})$. We need the following.

**1.6 Theorem** (TATE'S LOCAL DUALITY [BA]). *Let $k$ be a local field, $E$ an elliptic curve over $k$, $\alpha \in End(E/k)$, $\overline{\alpha}$ the dual isogeny.*

*(i) $E(k)$ and $H^1(G_k, E(\overline{k}))$ are Pontrijagin dual of each other (the first is compact, the second discrete).*

*(ii) The duality induces duality between $E(k)/\overline{\alpha}E(k)$ and $H^1(G_k, E(\overline{k}))[\alpha]$.*

*(iii) If $k'$ is a finite extension of $k$, then the transpose of $N_{k'/k} : E(k') \to E(k)$ is $Res : H^1(G_k, E(\overline{k})) \to H^1(G_{k'}, E(\overline{k}))$.*

To pair $u \in E(k)/mE(k)$ with $v \in H^1(G_k, E(\overline{k}))[m]$, one lets $u'$ be the image of $u$ in $H^1(G_k, E[m])$, and $v'$ any lifting of $v$ there (cf. (7)). Then take $u' \cup v' \in H^2(G_k, \mu_m) \hookrightarrow Br(k) \cong \mathbf{Q}/\mathbf{Z}$, using the Weil pairing, and the canonical isomorphism of the Brauer group with $\mathbf{Q}/\mathbf{Z}$.

**1.7** To conclude the proof of 1.5 let $\alpha \in \mathcal{O}_K$ be such that $(\alpha) = \mathfrak{p}^h$. Then, setting $k_n$ = the completion of $F_n$ at $P$, $H^1(G_P, E(\overline{F}_P))[\alpha]$ is dual, in view of 1.6(ii) and (iii), to

$$(13) \qquad \varprojlim E(k_n)/\overline{\alpha}E(k_n),$$

inverse limit with respect to the norm maps. However, since $(\overline{\alpha}, P) = 1$, and $P$ is a place of good reduction, $\overline{\alpha}$ induces an *isomorphism* on $E_1(k_n)$, the kernel of reduction *mod* $P$. Thus (13) is equal to $\varprojlim \tilde{E}(\kappa_n)/\overline{\alpha}\tilde{E}(\kappa_n)$, where $\tilde{E}$ is the reduced curve and $\kappa_n$ the residue field of $k_n$. Now $P$ is totally ramified in $F_\infty/F$ (cf. II.1.9(i)), so $\kappa_n = \kappa_0$ for all $n$, and $\tilde{E}(\kappa_0)/\overline{\alpha}\tilde{E}(\kappa_0)$ is a fixed finite $p$-group. The reductions of the norm maps become multiplication by $p^{m-n}$ (for $N_{m,n}$, $m \geq n \geq 1$), so the inverse limit vanishes. By duality, $H^1(G_P, E(\overline{F}_P))[\alpha] = 0$.

**1.8** Here is an example of the usefulness of theorem 1.5.

**Corollary.** *Suppose $F(E_{tor})$ is an abelian extension of $K$ (cf. II.1.4(12)). Then $E(F_\infty)$ is finitely generated modulo its torsion.*

PROOF: In this case $\mathcal{X}(F_\infty)$ is a finitely generated free $\mathbf{Z}_p$ module (III 1.7, 1.13 and 2.12). From the theorem $S(E/F_\infty)(\mathfrak{p}) \cong (\mathbf{Q}_p/\mathbf{Z}_p)^r$ for some $r$. The corollary follows from the descent sequence (12), with $F_\infty$ replacing $F$, because it is easy to see that $E(F_\infty)/torsion$ is a *free* abelian group.

## 2. THE THEOREM OF COATES-WILES

In this section we shall prove the following.

**2.1 Theorem.** *Let $K$ be a quadratic imaginary field, $F$ a finite abelian extension of $K$, and $E$ an elliptic curve over $F$ with complex multiplication by $K$. Assume that $F(E_{tor})$ is an abelian extension of $K$. If $E(F)$ is infinite, then $L(E/F, 1) = 0$.*

The method of proof is $p$-adic. The idea is to show that the "algebraic part" of the special value $L(\overline{\psi}, 1)$ ($\psi = \psi_{E/F}$ is the grossencharacter of $E$ over $F$) is divisible by arbitrarily high powers of a chosen prime. This is a variation (due to K. Rubin) on the original proof, which showed that the value in question is divisible by infinitely many primes, another property peculiar to zero!

**2.2** We may assume, to begin with, that $E$ is given in Weierstrass form. Let $\varphi$ be a grossencharacter of $K$ satisfying $\psi_{E/F} = \varphi \circ N_{F/K}$ (II.1.4), and $\mathfrak{f} = \ell.c.m.(\mathfrak{f}_\varphi, \mathfrak{f}_{F/K})$. Changing $E$ by isogeny, if necessary, we may assume that it has complex multiplication by $\mathcal{O}_K$, and that the lattice of periods of $\omega_E = \dfrac{dx}{y}$ is $L = \mathfrak{f}\Omega$ for some $\Omega \in \mathbf{C}^\times$. All this is standard notation from chapter II.

Choose an auxiliary prime $p$ which splits completely in $F$, and is relatively prime to $6\mathfrak{f}\Delta(L)$. Let $\mathfrak{p}$ be one of the factors of $p$ in $K$, and embed $\overline{\mathbf{Q}}$ once and for all in $\mathbf{C}_p$ so that the place induced on $K$ is $\mathfrak{p}$. For any prime $P$ of $F$ above $\mathfrak{p}$, $\pi = \psi(P) = \varphi(\mathfrak{p})$ is a generator of $\mathfrak{p}$, which is therefore principal.

**2.3** Suppose $a \in E(F)$ is a point of infinite order. Replacing it by some multiple, if necessary, we may assume that $a$ is in the kernel of reduction *mod* $P$ for each $P|\mathfrak{p}$, hence $|t(a)|_P < 1$ $\left(t = -\frac{2x}{y}\text{as usual}\right)$.

Let $F_n = F(E[\mathfrak{p}^n])$. $K(\mathfrak{f})$ and $F_n$ are linearly disjoint over $F$, and $K(\mathfrak{f})F_n = K(\mathfrak{f}\mathfrak{p}^n)$ (cf. II.1.7). Each of the $[F : K]$ primes $P$ above $\mathfrak{p}$ is totally ramified in $F_\infty$ (II.1.9), and we denote by the same letter $P$ the unique prime of $F_n$ above it, $0 \leq n \leq \infty$.

Let $a_n \in E(\overline{F})$ be a solution of

$$\pi^n(a_n) = a, \tag{1}$$

and $L_n = F_n(a_n)$. The *Kummer map*

$$Gal(L_n/F_n) \to E[\mathfrak{p}^n], \qquad \sigma \mapsto \sigma(a_n) - a_n, \tag{2}$$

exhibits $Gal(L_n/F_n)$ as a subgroup of the $\mathfrak{p}^n$ torsion points. Since $E$ has good reduction everywhere over $F_n$ (proof of theorem 1.5, $n \geq 1$), $L_n/F_n$ is unramified outside $\mathfrak{p}$.

**2.4** Fix a generator of the Tate module of $E[\mathfrak{p}^\infty]$ : $u_n \in E[\mathfrak{p}^n]$, $u_n \notin E[\mathfrak{p}^{n-1}]$, $\pi(u_n) = u_{n-1}$. Let $\omega_n = t(u_n)$ be the value of the local parameter $t = -\frac{2x}{y}$ at $u_n$. Then $\omega_n \in F_n$ is of valuation 1 at each $P|\mathfrak{p}$.

Let $e = (e_n)$ be a norm-compatible sequence of units in the tower $(F_n)$. Let $R$ be the completion of $\mathcal{O}_F$ at $\mathfrak{p}$ (II.4.1(2)), and

$$g_e(T) \in R[[T]]^x \tag{3}$$

the Coleman power series of $e$. (See II.4.5).

The idelic Artin symbol $(e_n, L_n/F_n)$ is trivial, since $e_n$ is a principal idele. On the other hand, it is equal to the sum of the local Artin symbols, and there only primes above $\mathfrak{p}$ contribute, since $L_n/F_n$ is unramified outside $\mathfrak{p}$, $e_n$ is a unit, and $F_n$ is totally complex. Therefore

$$0 = \sum_{P|\mathfrak{p}} \{(e_n, L_{n,P}/F_{n,P})(a_n) - a_n\}. \tag{4}$$

**2.5** Let $\lambda$ be the logarithm of the formal group $\hat{E}$ of $E$, with respect to the parameter $t$. It is a power series with coefficients in $F$, which we also consider in $F \otimes_K K_{\mathfrak{p}}$. Let $D_0 = \frac{1}{\lambda'(T)} \frac{d}{dT}$ be the translation invariant derivation on $\hat{E}$. Recall that the explicit reciptocity law I.4.2 computes for us the local contributions to (4) as follows.

$$t((e_n, L_{n,P}/F_{n,P})(a_n) - a_n) = \tag{5}$$
$$[\pi^{-n} Tr_n\{\lambda(t(a))\cdot D_0 \log g_e(\omega_n)\}]_P(\omega_n).$$

Here $Tr_n$ is the trace from $F_n$ to $F$, which is the same as the trace from $F_{n,P}$ to $F_P$. The subscript $P$ is meant to remind us that we deal with the formal group $\hat{E}$ over $\mathcal{O}_P$. Now $\lambda(t(a))$ can be pulled out of the trace because $a \in E(F)$. When we do so, the formalism of $g_e(T)$, and in particular property I.2.1(iii), (see also II.4.5(iii)) give for (5) the value

$$\left[(1 - \frac{1}{\pi})\cdot\lambda(t(a))\cdot D_0 \log g_e(0)\right]_P (\omega_n). \tag{6}$$

As $Gal(F/K)$ acts transitively on the $P$'s, (4), (5) and (6) give

$$\left(1 - \frac{1}{\pi}\right) Tr_{F/K}\{\lambda(t(a))\cdot D_0 \log g_e(0)\} \equiv 0 \bmod \pi^n \tag{7}$$

for each $n \geq 0$. Here we view $g_e(T)$ and $\lambda(T)$ as power series over $F \otimes_K K_{\mathfrak{p}} = \Phi$, and $Tr_{F/K}$ is the trace from $\Phi$ to $K_{\mathfrak{p}}$. Since $n$ is arbitrary,

$$\sum_{\mathfrak{c}} \{\lambda(t(a))\cdot D_0 \log g_e(0)\}^{\sigma_{\mathfrak{c}}} = 0, \tag{8}$$

where $\mathfrak{c}$ runs over ideals of $K$ whose Artin symbols represent $Gal(F/K)$.

**2.6** Let $\Omega_p$ be the $p$-adic period associated with $\Omega$ as in chapter II.4. Recall (II.4.3(10)) that $\Omega_p^{\sigma_{\mathfrak{c}}-1} = \Lambda(\mathfrak{c})\mathbf{N}\mathfrak{c}^{-1}$. Recall also (II.4.7(18)) that if we put $D = \Omega_p D_0$

$$\sigma_{\mathfrak{c}}(D \log g_e(0)) = \sigma_{\mathfrak{c}}(\delta_1(e)) = \mathbf{N}\mathfrak{c}^{-1}\, \delta_1(\sigma_{\mathfrak{c}}(e)), \tag{9}$$

where $\delta_1$ is the first Kummer logarithmic derivative. "Removing the Euler factor at $\mathfrak{p}$" gives (II.4.7(17))

$$\left(1 - \frac{\pi}{p}\right) \cdot \sigma_{\mathfrak{c}}(D \log g_e(0)) = \mathbf{N}\mathfrak{c}^{-1} \cdot \tilde{\delta}_1(\sigma_{\mathfrak{c}}(e))$$

$$(10) \qquad = \mathbf{N}\mathfrak{c}^{-1} \cdot \int_G \varphi(\sigma) d\mu_{\sigma_{\mathfrak{c}}(e)}(\sigma) \quad (G = Gal(F_\infty/F))$$

$$= \mathbf{N}\mathfrak{c}^{-1} \cdot \int_{\sigma_{\mathfrak{c}}^{-1} G} \varphi(\sigma_{\mathfrak{c}}\sigma) d\mu_e(\sigma).$$

Here we have used various elementary properties of $\tilde{\delta}_1(\beta)$ and $\mu_\beta$. See I.3.4-3.5. To sum up, (8) is equivalent to

$$(11) \qquad \sum_{\mathfrak{c}} \left[\lambda(t(a))^{\sigma_{\mathfrak{c}}} \cdot \frac{\varphi(\mathfrak{c})}{\Lambda(\mathfrak{c})}\right] \left[\int_{\sigma_{\mathfrak{c}}^{-1} G} \varphi(\sigma) d\mu_e(\sigma)\right] = 0.$$

In the last sum, the two quantities in square brackets depend only on $(\mathfrak{c}, F/K)$. We denote them, for brevity,

$$(12) \qquad W(\mathfrak{c}) = \lambda(t(a))^{\sigma_{\mathfrak{c}}} \cdot \frac{\varphi(\mathfrak{c})}{\Lambda(\mathfrak{c})},$$

$$(13) \qquad M(\mathfrak{c}) = \int_{\sigma_{\mathfrak{c}}^{-1} G} \varphi(\sigma) d\mu_e(\sigma).$$

**2.7** We now claim that, in fact, for any $(\mathfrak{d}, \mathfrak{f}) = 1$,

$$(14) \qquad \sum_{\mathfrak{c}} W(\mathfrak{c}) M(\mathfrak{c}\mathfrak{d}^{-1}) = 0.$$

Alternatively, $\sum_{\mathfrak{c}} W(\mathfrak{c}\mathfrak{d}) M(\mathfrak{c}) = 0$. Indeed,

$$W(\mathfrak{c}\mathfrak{d}) = \left(\lambda(t(a))^{\sigma_{\mathfrak{c}\mathfrak{d}}} \cdot \frac{\varphi(\mathfrak{c})}{\Lambda(\mathfrak{c})^{\sigma_{\mathfrak{d}}}}\right) \cdot \frac{\varphi(\mathfrak{d})}{\Lambda(\mathfrak{d})},$$

so up to a constant $\varphi(\mathfrak{d})/\Lambda(\mathfrak{d})$, independent of $\mathfrak{c}$, $W(\mathfrak{c}\mathfrak{d})$ is the same as $W(\mathfrak{c})$, provided we replace the elliptic curve $E$ with which we started by its conjugate $E^{\sigma_{\mathfrak{d}}}$, and $a$ by $\sigma_{\mathfrak{d}}(a)$. Note that the measure $\mu_e$ is independent of $E$.

Consider the cyclic matrix $(M(\mathfrak{c}\mathfrak{d}^{-1})) = M$ whose rows and columns are labeled by $Gal(F/K)$. *Since $a$ is of infinite order,* $W(\mathfrak{c}) \neq 0$, hence $M$ is singular, and $det(M) = 0$. The *Frobenius determinant* relation ([La3], p. 89) yields

$$\prod_{\chi \in \widehat{Gal(F/K)}} \left(\sum \chi(\mathfrak{c}) M(\mathfrak{c})\right) = 0. \tag{15}$$

**2.8** We now specify $e$. Let $e = (e_n(\mathfrak{a}))$, where

$$e'_n(\mathfrak{a}) = \Theta(1; \mathfrak{f}\mathfrak{p}^n, \mathfrak{a}) \quad \text{(see II.4.9(23))}, \tag{16}$$

$$e_n(\mathfrak{a}) = N_{K(\mathfrak{f}\mathfrak{p}^n)/F_n}(e'_n(\mathfrak{a})). \tag{17}$$

The measures $\mu_e$ on $\mathcal{G} = Gal(F_\infty/K)$ are then given by (II.4.12)

$$\mu_e = 12(\sigma_{\mathfrak{a}} - \mathbf{N}\mathfrak{a})\mu \tag{18}$$

where $\mu$ is the measure induced on $\mathcal{G}$ from $\mu(\mathfrak{f})$ via the natural projection $\mathbf{D}[[\mathcal{G}(\mathfrak{f})]] \to \mathbf{D}[[\mathcal{G}]]$. We easily deduce from (15) that

$$\prod_{\chi} \int_{\mathcal{G}} \varphi\chi(\sigma) d\mu(\sigma) = 0. \tag{19}$$

The interpolation formula II.4.2(31) gives, on the complex side,

$$\prod_{\chi} L(\overline{\chi\varphi}, 1) = 0. \tag{20}$$

Theorem 2.1 follows from this since $\prod L(\overline{\chi\varphi}, 1) = L(\overline{\psi}, 1)$, and $L(E/F, 1) = L(\psi, 1)L(\overline{\psi}, 1)$. The proof is now complete.

**2.9** REMARKS: When $F = K$ the proof is shortened considerably, because in (8) there is only one term and $\lambda(t(a))$ can be dropped. The use of the Frobenius determinant in 2.6-2.8 to complete the proof in the general case seems to be new. In [Ru1] a result of Bertrand in transcendental $p$-adic number theory is used instead. However, Rubin succeeds in saying which of the $L(\varphi\chi, 1)$ vanishes. For this we need to be able to distinguish between the various $\varphi\chi$, $\chi \in \widehat{Gal(F/K)}$, and this is possible if we introduce the abelian variety $Res_{F/K}E$, which is lurking behind many of our arguments. To keep the exposition more elementary, and to avoid Bertrand's theorem, we contented ourselves with theorem 2.1.

Our proof is slick in the sense that the use of the explicit reciprocity law obscures the role played by descent. Indeed, we did not even have to know the definition of the Selmer group, let alone a result like theorem 1.5, or its analogues "at finite levels." However, this is not the way the theorem was discovered in the first place. Coates and Wiles resorted to techniques involving the explicit reciprocity law, only because in the descent sequence they could not tell whether a point of infinite order gave rise to an abelian extension of $K(E[\mathfrak{p}])$ which was *truly ramified* at $\mathfrak{p}$. The recent ideas of F. Thaine and K. Rubin enable one to revive the original approach, and free the proof from the explicit reciprocity law, at least when $F = K$.

## 3. GREENBERG'S THEOREM

Theorem 2.1 concludes the vanishing of the zeta function of $E$ at 1, from the existence of a point of infinite order. In the converse direction we have a theorem of R. Greenberg. Not as strong as the Coates-Wiles theorem, this result relies on the finiteness of the $p$-primary part of the Tate Shafarevitch group, for some prime $p$ of good ordinary reduction. It only applies to a more restricted class of elliptic curves. A much more definite result is contained in the work of Gross and Zagier [G-Z], at least when the field of definition is $\mathbf{Q}$. Nevertheless, Greenberg's theorem is a beautiful application of the ideas developed in chapters II and III, and was recently put to use, together with [G-Z] and other ingredients, to yield the following strengthening of theorem 2.1 (about which we say nothing more here).

**Theorem** (K. RUBIN [RU3]). *Suppose that $E$ is an elliptic curve defined over $\mathbf{Q}$, with complex multiplication by $K$. Then*

*(i)* $L(E/\mathbf{Q},1) \neq 0 \Rightarrow E(\mathbf{Q})$ *is finite.*

*(ii)* $L(E/\mathbf{Q},1) = 0$, $L'(E/\mathbf{Q},1) \neq 0 \Rightarrow rk\ E(\mathbf{Q}) = 1$.

**3.1** We begin by describing the class of elliptic curves with which we shall be dealing. Let $K$ be a quadratic imaginary field, and $F$ an abelian extension of $K$. $F$ is called an *anti-cyclotomic extension* of $K$ if it is Galois over $\mathbf{Q}$, and $Gal(K/\mathbf{Q})$ acts on $Gal(F/K)$ by $-1$. Thus if $\rho$ denotes complex conjugation and $\sigma \in Gal(F/K)$, $\rho\sigma\rho^{-1} = \sigma^{-1}$. A typical example is the Hilbert class field of $K$. Fix such an extension $F$, and set $F' = F \cap \mathbf{R}$.

Let $E$ be an elliptic curve defined over $F'$, which, over $F$, has complex multiplications by $\mathcal{O}_K$. Assume in addition:

(i) $F(E_{tor})$ is an abelian extension of $K$.

(ii) If $\varphi$ is a grossencharacter of $K$ for which $\varphi \circ N_{F/K} = \psi_{E/F}$ (see II.1.4), then

$$\varphi(\overline{\mathfrak{a}}) = \overline{\varphi}(\mathfrak{a}). \tag{1}$$

Since $\chi(\overline{\mathfrak{a}}) = \overline{\chi}(\mathfrak{a})$ for any $\chi \in \widehat{Gal(F/K)}$, if (1) holds with one choice of $\varphi$, it holds with any.

EXAMPLES: (a) If $d_K = q$ is a prime congruent to 3 *mod* 4, and $F$ is the Hilbert class field of $K$, the curves $A(q)^d$ ($d$ = *a* square free integer), introduced by Gross in [Gr] p. 35, satisfy (i) and (ii). In this case genus theory easily implies that the class number of $K$ is odd, so (ii) is superfluous, as the next example shows.

(b) If $[F : K]$ is odd, (ii) is a consequence of (i). Indeed, since $E$ is defined over $F'$, $\psi(\overline{\mathfrak{A}}) = \overline{\psi}(\mathfrak{A})$. Letting $c(\psi) = \rho \circ \psi \circ \rho^{-1}$, $c(\psi) = \psi$, hence $c(\varphi) \circ N_{F/K} = \varphi \circ N_{F/K}$, and $c(\varphi) = \varphi \cdot \chi$ for some $\chi \in \widehat{Gal(F/K)}$. Since $c(\chi) = \chi$, $\varphi = c(c(\varphi)) = \varphi\chi^2$, showing $\chi^2 = 1$, and eventually $\chi = 1$, because $[F : K]$ is odd.

(c) If $K$ has class number 1, and $F = K$, (i) and (ii) are automatically satisfied.

**3.2 Theorem** (CF. [GRE2]). *Let $F/K$ be an anticyclotomic extension, $F' = F \cap \mathbf{R}$, and $E$ an elliptic curve over $F'$, satisfying conditions (i) and (ii) above.*

*Suppose that for some $\chi \in Gal(\widehat{F}/K)$ $L(\overline{\varphi\chi}, s)$ has a zero of odd order at $s = 1$. Then either $E(F')$ is infinite, or $\text{Ш}(E/F')$ has an infinite p-primary component for every odd prime p that splits in K, is unramified in F, is relatively prime to $[F : K]$, and above which E has good reduction (in $F'$).*

Of course, the second alternative is believed not to occur. The hypothesis is clearly satisfied if $L(\overline{\psi}, s)$ itself has a zero of odd order at $s = 1$. If $F$ is the Hilbert class field of $K$, then, according to a theorem of Shimura ([Sh2]), the vanishing of $L(\overline{\varphi\chi}, 1)$ is a property independent of $\chi$. In that case, it is also known that $rk\ E(F')$ is a multiple of $[F : K] = h_K$ ([Gr] theorem 16.1.3).

The proof of theorem 3.2 is carried out in two main steps. In 3.3-3.9 we use descent and theorem 1.5 (relating the Selmer group to the Iwasawa module $\mathcal{X}$), to reduce it to a statement about the characteristic power series of $\mathcal{X}$ (10).

In 3.10-3.13 we prove this statement. The key ingredients are the theory of $p$-adic $L$ functions as developed in chapters II and III, and a non-vanishing theorem for *complex* $L$ functions.

**3.3** Pick a prime $p$ as in the statement of the theorem. In light of the descent exact sequence (see 1.3)

(2) $$0 \to E(F') \otimes \mathbf{Q}_p/\mathbf{Z}_p \to S(E/F')(p) \to \text{Ш}(E/F')(p) \to 0,$$

we actually have to prove that the $p$-primary part of the Selmer group over $F'$ is infinite.

We now describe the various fields that we shall encounter, and give them names. The notation is different from the one used in chapters II and III, because we need to consider more fields.

Let $\mathcal{F}_\infty = F(E[p^\infty])$, $\mathcal{F}_\infty(\mathfrak{p}) = F(E[\mathfrak{p}^\infty])$, and $\mathcal{F}_\infty(\overline{\mathfrak{p}}) = F(E[\overline{\mathfrak{p}}^\infty])$. Let $\mathcal{G}$, $\mathcal{G}_1$, $\mathcal{G}_2$ be the Galois groups of these fields over $K$, and $G$, $G_1$, $G_2$ the subgroups fixing $F$. Then, since $p \neq 2$ and $p$ is unramified in $F$, there are canonical isomorphisms

(3) $$\begin{cases} \kappa_1 : G_1 = \Delta_1 \times \Gamma_1 \simeq \mathcal{O}_{\mathfrak{p}}^x \simeq \mathbf{Z}_p^x, \\ \kappa_2 : G_2 = \Delta_2 \times \Gamma_2 \simeq \mathcal{O}_{\overline{\mathfrak{p}}}^x \simeq \mathbf{Z}_p^x. \end{cases}$$

As usual, $|\Delta_i| = p - 1$, $\Gamma_i = \kappa_i^{-1}(1 + p\mathbf{Z}_p)$, and we write $\Delta = \Delta_1 \times \Delta_2$, $\Gamma = \Gamma_1 \times \Gamma_2$.

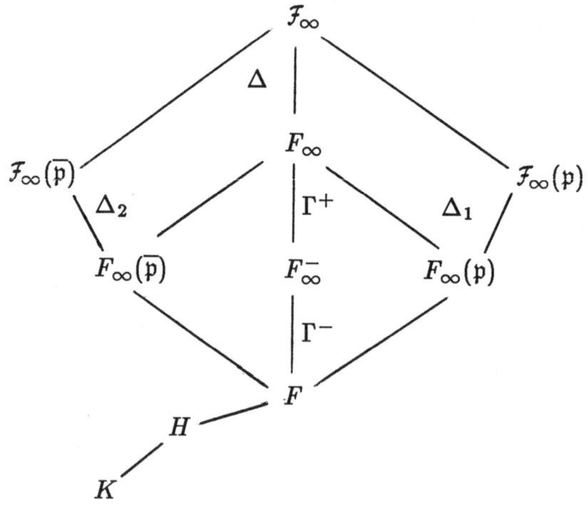


We let $F_\infty(\mathfrak{p})$ (resp. $F_\infty(\overline{\mathfrak{p}})$) be the fixed field of $\Delta_1$ (resp. $\Delta_2$) in $\mathcal{F}_\infty(\mathfrak{p})$ (resp. $\mathcal{F}_\infty(\overline{\mathfrak{p}})$), and $F_\infty = F_\infty(\mathfrak{p})F_\infty(\overline{\mathfrak{p}})$. The field $F_\infty$ contains the *cyclotomic* $\mathbf{Z}_p$ *extension* $F_\infty^+$ of $F$, as well as the *maximal anticyclotomic* extension of $K$ in $F_\infty$, which we denote $F_\infty^-$. The latter is the field fixed by $\rho\sigma\rho^{-1}\sigma$, for all $\sigma \in Gal(F_\infty/K)$ ($\rho$ is complex conjugation), and it contains $F$ by assumption. Since $p \neq 2$, $F_\infty^+$ and $F_\infty^-$ are linearly disjoint over $F$, and their compositum is $F_\infty$. If we now let $\kappa^+ = \kappa_1\kappa_2$, $\kappa^- = \kappa_1\kappa_2^{-1}$ ($\kappa_i$ is viewed as a character of $G$ after projecting $G$ to $G_i$), $\kappa^+$ is the cyclotomic character, and factors through $\mathcal{F}_\infty^+ = F(\mu_{p^\infty})$. This is a consequence of the Weil pairing. On the other hand $\kappa^-$ cuts the maximal anticyclotomic subextension $\mathcal{F}_\infty^-$ of $\mathcal{F}_\infty$. Note that $[\mathcal{F}_\infty^+ \cap \mathcal{F}_\infty^- : F] = 2 = [\mathcal{F}_\infty : \mathcal{F}_\infty^+\mathcal{F}_\infty^-]$, but restricted to $\Gamma$, $\kappa^-$ factors through $F_\infty^-$, and yields an isomorphism $Gal(F_\infty^-/F) = \Gamma^- \simeq 1 + p\mathbf{Z}_p$.

It is useful to fix a generator $u$ of $1 + p\mathbf{Z}_p$, and to define $\gamma_1$, $\gamma_2$, $\gamma^+$ and $\gamma^-$

in $\Gamma$ by

$$\text{(4)} \qquad \begin{cases} \kappa_1(\gamma_1) = u, & \kappa_1(\gamma_2) = 1, \\ \kappa_2(\gamma_1) = 1, & \kappa_2(\gamma_2) = u, \\ \gamma^+ = \gamma_1\gamma_2, & \gamma^- = \gamma_1/\gamma_2. \end{cases}$$

Then $\gamma_1$, $\gamma_2$, $\gamma^+$ and $\gamma^-$ project to generators of $\Gamma_1$, $\Gamma_2$, $\Gamma^+ = Gal(F_\infty^+/F)$ and $\Gamma^- = Gal(F_\infty^-/F)$, and are trivial on $\Gamma_2$, $\Gamma_1$, $\Gamma^-$ and $\Gamma^+$ respectively.

**3.4 Lemma.** *Notation as above, the following are equivalent.*

*(a) The conclusion of theorem 3.2.*

*(b) $S(E/F')(p)$ is infinite.*

*(c) $S(E/F)(\mathfrak{p})$ is infinite.*

PROOF: *(b)* $\Leftrightarrow$ *(a)*: Already observed in 3.3.

*(c)* $\Leftrightarrow$ *(b)*: The inflation-restriction sequence shows that $S(E/F')(p) = S(E/F)(p)^{Gal(F/F')}$ $(p \neq 2)$. Since $S(E/F)(p) = S(E/F)(\mathfrak{p}) \oplus S(E/F)(\overline{\mathfrak{p}})$, and complex conjugation interchanges these two factors, (c) is equivalent to (b).

**3.5 Lemma** ([PR1]). *The restriction homomorphism*

$$\text{(5)} \qquad S(E/F)(\mathfrak{p}) \xrightarrow{r} S(E/F_\infty^-)(\mathfrak{p})^{\Gamma^-}$$

*is injective, with finite cokernel.*

PROOF: Since $E(F_\infty^-)[\mathfrak{p}] = 0$, the inflation-restriction sequence yields an isomorphism

$$\text{(6)} \qquad H^1(Gal(\overline{F}/F), E[\mathfrak{p}^\infty]) \simeq H^1(Gal(\overline{F}/F_\infty^-), E[\mathfrak{p}^\infty])^{\Gamma^-},$$

hence (5) is clearly injective. To deal with $Coker(r)$, let us examine the local restriction maps (for each place $v$ of $\overline{F}$)

$$\text{(7)} \qquad H^1(F_v, E(\overline{F}_v)) \xrightarrow{r_v} H^1(F_{\infty,v}^-, E(\overline{F}_v)),$$

where we write $H^1(k,-)$ for $H^1(Gal(\overline{k}/k),-)$. We shall prove: (a) If $v \nmid p$, $r_v$ is injective. (b) If $v|p$, $Ker(r_v)$ is finite. The lemma follows from these two assertions, because by (6) the cokernel of $r$ maps injectively into $\amalg_v\ Ker(r_v)$.

PROOF OF (A): Assume first that $v$ is a place of good reduction. Since $F^-_{\infty,v}/F_v$ is *unramified*, every point in $E(F_v)$ is a norm from each $E(F^-_{n,v})$, so by Tate's local duality $Ker(r_v) = H^1(F^-_{\infty,v}/F_v, E(F^-_{\infty,v})) = 0$ (see [Maz2] corollary 4.4). If $v$ is a place of bad reduction, the same proof works, because we can "translate" the tower $F^-_{\infty}/F$ by the extension $\mathcal{F}_1(\mathfrak{p})$, which is of degree $p - 1$, and over which $E$ acquires good reduction everywhere. Having proven the same result there, we may descend to $F$ taking $\Delta_1$-invariants.

PROOF OF (B): A theorem of Mazur ([Maz2] proposition 4.39) implies that $Ker(r_v) = H^1(F^-_{\infty,v}/F_v, E(F^-_{\infty,v}))$ is finite.

**3.6 Lemma.** *The restriction homomorphism induces an isomorphism*

$$S(E/F^-_{\infty})(\mathfrak{p}) \simeq S(E/F_{\infty})(\mathfrak{p})^{\Gamma^+} \tag{8}$$

$(\Gamma^+ = Gal(F_\infty/F^-_\infty))$.

PROOF: As in the previous section, since $E(F_\infty)[\mathfrak{p}] = 0$, there is an isomorphism

$$H^1(Gal(\overline{F}/F^-_\infty), E[\mathfrak{p}^\infty]) \simeq H^1(Gal(\overline{F}/F_\infty), E[\mathfrak{p}^\infty])^{\Gamma^+}, \tag{9}$$

and (8) is injective. To prove the surjectivity, note that $F_\infty/F^-_\infty$ is *everywhere unramified*, so like claim (a) in 3.5, the local restriction maps $H^1(F^-_{\infty,v}, E(\overline{F}_v)) \xrightarrow{r_v} H^1(F_{\infty,v}, E(\overline{F}_v))$ are injective, as desired.

**3.7** We briefly recall the structure theory of $\mathbf{Z}_p[[\Gamma]]$-modules (recall $\Gamma = \Gamma_1 \times \Gamma_2 \cong \mathbf{Z}_p^2$), to which we alluded in III.1.14. Let $M$ be a noetherian torsion $\mathbf{Z}_p[[\Gamma]]$-module. $M$ is *pseudo-null* if its localization at any principal ideal ($\equiv$ ideal of height 1) of $\mathbf{Z}_p[[\Gamma]]$ is trivial. Geometrically, the associated coherent sheaf is supported in codimension 2. The notion of *pseudo-isomorphism*, and the *characteristic ideal char.*$(M)$ are defined precisely as in III.1.8. Thus there exists an exact sequence

$$0 \to N \to M \to \prod_{1 \le i \le r} \mathbf{Z}_p[[\Gamma]]/(f_i) \to N' \to 0$$

with $N$ and $N'$ pseudo-null, and $char.(M) = (\prod f_i)$. For details see [Cu] and Bourbaki's Commutative Algebra.

**Proposition.** *Let $\mathcal{X} = Gal(M(\mathcal{F}_\infty)/\mathcal{F}_\infty)$ be the Galois group of the maximal abelian p-extension of $\mathcal{F}_\infty = F(E[\mathfrak{p}^\infty])$ unramified outside $\mathfrak{p}$, over $\mathcal{F}_\infty$. Let $\omega = \kappa_1|\Delta : \Delta \to \mathbf{Z}_p^x$ (recall $\Delta = Gal(\mathcal{F}_\infty/F_\infty)$) be the character giving the action of $\Delta$ on $E[\mathfrak{p}]$, and $\mathcal{X}^\omega$ the corresponding eigenspace of $\mathcal{X}$. Let $f_\omega \in \mathbf{Z}_p[[\Gamma]]$ be a characteristic power series for $\mathcal{X}^\omega$ : $char.(\mathcal{X}^\omega) = (f_\omega)$. Let $\gamma^+$ be a generator of $\Gamma^+ = Gal(F_\infty/F_\infty^-)$ and suppose that*

(10) $$(\gamma^+ - \kappa_1(\gamma^+))|f_\omega.$$

*Then the Pontrijagin dual of $S(E/F_\infty^-)(\mathfrak{p})$ is not $\mathbf{Z}_p[[\Gamma^-]]$-torsion.*

**3.8** Before proving the proposition, let us draw the consequence that we need.

**Corollary.** *Suppose that (10) holds. Then $S(E/F)(\mathfrak{p})$ is infinite, hence by lemma 3.4 the conclusion of theorem 3.2 holds too.*

PROOF: Let $Y$ be the Pontrijagin dual of $S(E/F_\infty^-)(\mathfrak{p})$. Then the dual of the $\Gamma^-$-invariants is $Y/(\gamma^- - 1)Y$, where $\gamma^-$ is a generator of $\Gamma^-$. If this last group is finite, say of order $p^m$, then $p^m(\gamma^- - 1)$ annihilates $Y$, contrary to proposition 3.7. Thus $Y/(\gamma^- - 1)Y$, as well as $S(E/F_\infty^-)(\mathfrak{p})^{\Gamma^-}$, are infinite. Lemma 3.5 concludes the proof.

**3.9** PROOF OF PROPOSITION 3.7: By theorem 1.5

$$S(E/F_\infty)(\mathfrak{p}) \simeq Hom(\mathcal{X}, E[\mathfrak{p}^\infty])^\Delta.$$

Combining this with lemma 3.6,

$$(11)\qquad S(E/F_{\infty}^{-})(\mathfrak{p}) \simeq Hom(\mathcal{X}, E[\mathfrak{p}^{\infty}])^{\Delta \times \Gamma^{+}}$$
$$= Hom(\mathcal{X}^{\omega}/(\gamma^{+} - \kappa_1(\gamma^{+}))\mathcal{X}^{\omega}, E[\mathfrak{p}^{\infty}]).$$

The module $\mathcal{X}$ is a noetherian torsion $\mathbf{Z}_p[[\Gamma]]$-module, as is implicitly included in the statement of proposition 3.7. This is proved by the same argument used for the "one-variable" module at the end of III.1.7. By the structure theory for $\mathbf{Z}_p[[\Gamma]]$-modules, there exists an exact sequence

$$(12)\qquad 0 \to N \to \mathcal{X}^{\omega} \to \prod_{1 \le i \le r} \mathbf{Z}_p[[\Gamma]]/(f_i) \to N' \to 0$$

where $N$ and $N'$ are pseudo-null, and $f_\omega = \prod f_i$. By our assumption $\gamma^{+} - \kappa_1(\gamma^{+})$ divides one of the $f_i$, say $f_1$. Now, if $N$ is pseudo-null, it is annihilated by two relatively prime elements of $\mathbf{Z}_p[[\Gamma]]$, hence $N/(\gamma^{+} - \kappa_1(\gamma^{+}))N$ is still $\mathbf{Z}_p[[\Gamma^{-}]]$-torsion. Since $\mathbf{Z}_p[[\Gamma]]/(f_1, \gamma^{+} - \kappa_1(\gamma^{+})) \cong \mathbf{Z}_p[[\Gamma^{-}]]$, $\mathcal{X}^{\omega}/(\gamma^{+} - \kappa_1(\gamma^{+}))\mathcal{X}^{\omega}$ can not be $\mathbf{Z}_p[[\Gamma^{-}]]$-torsion, and the proposition follows from (11).

**3.10** ROOT NUMBERS: We begin the second part of the proof of theorem 3.2 with a discussion of root numbers. Without loss of generality, we may assume that $\varphi$ is chosen such that $L(\overline{\varphi}, s)$ has a zero of odd order at $s = 1$. Let

$$(13)\qquad \varepsilon_k = \overline{\varphi}^{2k+1}\mathbf{N}^{-k-1} = \overline{\varphi}^{k}\varphi^{-k-1}, \quad k \ge 0.$$

Since $\varphi(\overline{\mathfrak{a}}) = \overline{\varphi}(\mathfrak{a})$ by assumption, $\check{\varepsilon}_k = \varepsilon_k$ (i.e. $\varepsilon_k$ is an *anticyclotomic* character (II.6.5)). Since the functional equation relates $L(\varepsilon_k, s)$ to $L(\varepsilon_k, -s)$, $W(\varepsilon_k) = \pm 1$ (II.6.1(3)). By assumption, $W(\varepsilon_0) = -1$.

**Lemma.** *Let $m = (p-1)[F:K]$. Then*

*(i) If $k, j \ge 0$, $k \equiv j \bmod m$, then $W(\varepsilon_k) = W(\varepsilon_j)$.*

*(ii) $W(\varepsilon_{m-1}) = 1$.*

PROOF: Notice first that $(\varphi/\overline{\varphi})^m = \varphi^{2m}\mathbf{N}^{-m}$ is an unramified character. Indeed, letting $d = [F:K]$, $\varphi^d(\mathfrak{a}) = \varphi \circ N_{F/K}(\mathfrak{a}\mathcal{O}_F) = \psi(\mathfrak{a}\mathcal{O}_F) \in K^x$, so $\varphi^{dw_K}$ is unramified, and $w_K | p - 1$. Now quite in general, if $\lambda$ is a Hecke character of $K$ of infinity type $(k, j)$, put $\nu(\lambda) = |k - j|$. Then if $\lambda_1$ and $\lambda_2$ have *relatively prime* conductors $\mathfrak{f}_1$, $\mathfrak{f}_2$, we have for $\lambda = \lambda_1 \lambda_2$

$$(14) \qquad W(\lambda) = W(\lambda_1) W(\lambda_2) \tilde{\lambda}_1(\mathfrak{f}_2) \tilde{\lambda}_2(\mathfrak{f}_1)\, i^{\nu(\lambda_1)+\nu(\lambda_2)-\nu(\lambda)},$$

where $\tilde{\lambda} = \lambda/|\lambda|$. This follows from Tate's thesis, and may be found for example in [We3] p. 161, or deduced directly from II.6.1. Since we have seen that $\varphi^m$ is unramified, applying (14) to $\lambda_1 = \lambda_2 = \varphi^m$ we get $W((\varphi/\overline{\varphi})^m) = W(\varphi^{2m}) = W(\varphi^m)^2 = (\pm 1)^2 = 1$. Applying it with $\lambda_1 = \varepsilon_{k+m}$ and $\lambda_2 = (\varphi/\overline{\varphi})^m$, and using the fact that $\overline{\mathfrak{f}}_1 = \mathfrak{f}_1$ and that $\overline{\varphi}(\mathfrak{a}) = \varphi(\overline{\mathfrak{a}})$, we see that

$$W(\varepsilon_k) = W(\varepsilon_{k+m}) \cdot i^{|2k+2m+1|+|2m|-|2k+1|} = W(\varepsilon_{k+m}),$$

which proves (i). For (ii),

$$\begin{aligned} W(\varepsilon_{m-1}) &= W(\overline{\varphi}^{-1}(\overline{\varphi}/\varphi)^m) = W(\overline{\varphi}^{-1}) i^{1+2m-(2m-1)} \\ &= -W(\overline{\varphi}^{-1}) = -W(\varphi^{-1}) = 1. \end{aligned}$$

**3.11** THE NON-VANISHING THEOREM: The key to the rest of the proof is a result saying that *generically*, the numbers $L(\varepsilon_k, 0)$ do not vanish, unless they are forced to vanish by the sign in the functional equation. Recall that $m = (p - 1)[F : K]$.

**Theorem** ([GRE3] THEOREM 1, [ROH] P. 384). *If* $W(\varepsilon_k) = 1$, *then* $L(\varepsilon_j, 0) = 0$ *for only finitely many* $j \geq 0$, $j \equiv k$ *mod* $m$.

Indeed, by the previous lemma $W(\varepsilon_j) = 1$ for all those $j$. There are two proofs of the theorem. Rohrlich's proof (although stated only for $K$ with class number 1, it works in general) is purely complex-analytic, but uses a *non-archimedean* version of Roth's theorem at a crucial point. Greenberg uses a mixture of $p$-adic and complex arguments, accompanied by the *classical* version of Roth's theorem! Both yield stronger results than what is quoted above. Since their methods are different from the spirit of this book, and since the two papers are easily accessible, we omit the proof.

**3.12** Recall that $p$ did not divide $[F : K]$, hence

$$\mathcal{G} = Gal(\mathcal{F}_\infty/K) = \Gamma \times H,$$

where $H = Gal(\mathcal{F}_1/K)$ is of order prime to $p$, and $\Gamma = Gal(\mathcal{F}_\infty/\mathcal{F}_1) \cong \mathbf{Z}_p^2$. We may decompose any $p$-adic character $\varepsilon$ of $\mathcal{G}$ as $\varepsilon_\Gamma \varepsilon_H$, where $\varepsilon_\Gamma$ is trivial on $H$, and vice versa. In particular we have $\varphi = \varphi_\Gamma \varphi_H$, and $\varphi_H|\Delta = \kappa_1|\Delta = \omega$.

Complex conjugation $\rho$ acts on $\mathcal{G}$, and we let

$$c(\sigma) = \rho \circ \sigma \circ \rho^{-1}, \quad \sigma \in \mathcal{G}. \tag{15}$$

If $\varepsilon$ is a $p$-adic character of $\mathcal{G}$, so is $\varepsilon \circ c$. If $\varepsilon$ is a grossencharacter, $\varepsilon \circ c(\mathfrak{a}) = \varepsilon(\overline{\mathfrak{a}})$, so by our assumptions $\varphi \circ c = \overline{\varphi}$.

If $X$ is any $\mathbf{Z}_p[[\mathcal{G}]]$ module, and $\chi \in \hat{H}$, we write $X^\chi$ for $(\mathbf{D} \otimes_{\mathbf{Z}_p} X)^\chi$, where $\mathbf{D}$ is the ring of integers in the completion of $\mathbf{Q}_p^{ur}$. This slight abuse of notation is justified since we are only interested in $char.(X^\chi)$, and the characteristic ideal of a $\Gamma$-module behaves well under extension of scalars.

**Proposition.** *Fix a congruence class $k_0$ mod $m$. Let $\chi \in \hat{H}$ be defined by*

$$\chi = \varphi_H^{k_0+1} \overline{\varphi}_H^{-k_0}. \tag{16}$$

*Let $\mathfrak{g}$ be the prime-to-p part of $\mathfrak{f}_\chi$, and*

$$g_\chi = \chi(\mu(\mathfrak{g}\overline{\mathfrak{p}}^\infty)) \in \Lambda = \mathbf{D}[[\Gamma]] \tag{17}$$

*the primitive, "two-variable", p-adic $L$ function of $\chi$ (see II.4.14, III.1.10, 1.14). Then the following are equivalent:*

*(a) $L(\varepsilon_k, 0) = 0$ for infinitely many $k \geq 0$, $k \equiv k_0$ mod $m$.*

*(b) $L(\varepsilon_k, 0) = 0$ for every $k \geq 0$, $k \equiv k_0$ mod $m$.*

*(c) $(\gamma^+ - \kappa_1(\gamma^+))|g_\chi$, where $\gamma^+$ is a generator of $\Gamma^+ = Gal(F_\infty/F_\infty^-)$.*

PROOF: Let $T = \gamma^+ - \kappa_1(\gamma^+)$, $S = \gamma^- - 1$. It is well known that $\mathbf{Z}_p[[\Gamma]] = \mathbf{Z}_p[[S, T]]$, the power series ring in two variables (although a more standard choice for $T$ is $\gamma^+ - 1$). Furthermore, $\varepsilon_k^{-1}(T) = \kappa_1^{k+1}\kappa_2^{-k}(\gamma^+) - \kappa_1(\gamma^+) = \kappa_1(\gamma^+)(\kappa^-(\gamma^+)^k - 1) = 0$, because $\varphi_\Gamma = \kappa_1|\Gamma$, $\overline{\varphi}_\Gamma = \kappa_2|\Gamma$ (see (4)). Similarly, $\varepsilon_k^{-1}(S) = \kappa_1^{k+1}\kappa_2^{-k}(\gamma^-) - 1 = u^{2k+1} - 1$, where $u$ is a generator of $1 + p\mathbf{Z}_p$. It follows that if we write

$$g_\chi = a(S) + T\,b(S,T),$$

then for $k \equiv k_0 \bmod m$ $(\varepsilon_k^{-1})_H = \chi$, and

$$\int_{\mathcal{G}} \varepsilon_k^{-1}(\sigma)d\mu(\mathfrak{g}\overline{\mathfrak{p}}^{\infty};\sigma) = \int_\Gamma \varepsilon_k^{-1}(\sigma)dg_\chi(\sigma) \tag{18}$$
$$= g_\chi(u^{2k+1} - 1,0) = a(u^{2k+1} - 1).$$

We have identified the measure $g_\chi$ with the corresponding power series $g_\chi(S,T)$. Therefore $T$ divides $g_\chi$ if and only if (18) vanishes for infinitely many $k$, and in that case, it vanishes for all $k \equiv k_0 \bmod m$.

It remains to relate (18) to the special value of the complex $L$ function. However, this is precisely the contents of theorem II.4.14. Formula (36) there gives for (18)

$$\Omega_p^{-2k-1}\int_{\mathcal{G}} \varepsilon_k^{-1}\, d\mu(\mathfrak{g}\overline{\mathfrak{p}}^{\infty}) = \tag{19}$$
$$\Omega^{-2k-1}\left(\frac{\sqrt{d_K}}{2\pi}\right)^{-k}\left(1 - \frac{\varphi^{2k+1}(\mathfrak{p})}{p^{k+1}}\right)\cdot k!\cdot L(\varepsilon_k,0).$$

The proof of the proposition is now complete.

**3.13** We can now conclude the proof of theorem 3.2. Suppose that $p$ is as in the statement of that theorem, and consider the fundamental exact sequence III.1.7(13)

$$0 \to \mathcal{E}/\mathcal{C} \to \mathcal{U}/\mathcal{C} \to \mathcal{X} \to \mathcal{W} \to 0 \tag{20}$$

for the field $\mathcal{F}_\infty = F(E[p^\infty])$. Recall that $\mathcal{X}$ is the Galois group of the maximal abelian $p$ extension of $\mathcal{F}_\infty$ unramified outside $\mathfrak{p}$, $\mathcal{W}$ its (absolutely) unramified quotient, and $\mathcal{U}$, $\mathcal{E}$ and $\mathcal{C}$ the Iwasawa modules of local (at $\mathfrak{p}$), global, and elliptic units, respectively. More precisely, to get (20) from III.1.7(13) let $\mathfrak{f} = \ell.c.m.(\mathfrak{f}_{F/K}, \mathfrak{f}_\varphi)$, so that $F \subset K(\mathfrak{f})$. Take the inverse limit of III.1.7(13), with $\mathfrak{f}$ replaced by $\mathfrak{f}\overline{\mathfrak{p}}^m$, as $m \to \infty$. Finally take $Gal(K(\mathfrak{f})/F)$ invariants to descend from $K(\mathfrak{f}p^\infty)$ to $\mathcal{F}_\infty$.

We decompose the modules in (20) with respect to $\chi \in \hat{H}$, and call the characteristic power series of the resulting $\Lambda = \mathbf{D}[[\Gamma]]$-modules

$$h_\chi = char.(\mathcal{E}/\mathcal{C})^\chi$$

$$(21) \qquad g_\chi = char.(\mathcal{U}/\mathcal{C})^\chi$$

$$f_\chi = char.\mathcal{X}^\chi.$$

This notation is in accordance with (17), because of lemma III.1.10, which gives the $p$-adic $L$ function as the characteristic power series of $\mathcal{U}/\mathcal{C}$. Notice that the exceptional case $\chi = 1$ does not occur in the "two-variable" theory, because the $p$-adic $L$ function always has the $\overline{\mathfrak{p}}$-Euler factor removed from it. The notation also agrees with that of proposition 3.7 if we observe that for any $\theta \in \hat{\Delta}$

$$(22) \qquad \mathcal{X}^\theta = \prod_{\chi|\Delta=\theta} \mathcal{X}^\chi,$$

hence $f_\omega = \prod f_\chi$, the product taken over all the characters of $H$ extending $\omega$.

Recall that $\Lambda$ is a unique factorization domain. Clearly $h_\chi | g_\chi$, and $g_\chi | h_\chi f_\chi$. Now consider the special eigenspaces corresponding to $\chi = \varphi_H$, and to $\chi\circ c = \overline{\varphi}_H$. The module of global units modulo elliptic units in the field $K(\mathfrak{f}p^\infty)$ can be defined as the inverse limit of the $p$-Sylow subgroups of groups of global units modulo elliptic units in the fields $K(\mathfrak{f}p^n)$. These fields are Galois over $\mathbf{Q}$ ($\mathfrak{f} = \ell.c.m.(\mathfrak{f}_{F/K}, \mathfrak{f}_\varphi) = \overline{\mathfrak{f}}$), and complex conjugation preserves global or elliptic units in them. When we take $Gal(K(\mathfrak{f}p^\infty)/\mathcal{F}_\infty)$-invariants this is still true (note that $\mathcal{F}_\infty$ is also Galois over $\mathbf{Q}$). It follows that complex conjugation induces a natural isomorphism of $(\mathcal{E}/\mathcal{C})^\chi$ with $(\mathcal{E}/\mathcal{C})^{\chi\circ c}$ as groups. As $\Gamma$-modules we have, by "transport of structure", that

$$(23) \qquad h_{\chi\circ c} = c(h_\chi).$$

This is a crucial observation. It is *false* for $g_\chi$ or $f_\chi$, but, as expected from the main conjecture, true for $char.(\mathcal{W})$ at the other end of (20) (we do not need this fact).

Now $c(\gamma^+ - \kappa_1(\gamma^+)) = \gamma^+ - \kappa_1(\gamma^+)$, so

$$(24) \qquad (\gamma^+ - \kappa_1(\gamma^+))|h_\chi \Leftrightarrow (\gamma^+ - \kappa_1(\gamma^+))|h_{\chi^\circ c}.$$

According to our assumptions on $\varphi$ and lemma 3.10(i), $W(\varepsilon_k) = -1$ if $k \geq 0$, $k \equiv 0 \ mod \ m$, hence $L(\varepsilon_k, 0) = 0$. By proposition 3.12 ($k_0 = 0$),

$$(25) \qquad (\gamma^+ - \kappa_1(\gamma^+))|g_\chi.$$

On the other hand, by lemma 3.10(ii), $W(\varepsilon_k) = 1$ if $k \geq 0$, $k \equiv -1 \ mod \ m$, hence by the non-vanishing theorem (3.11), $L(\varepsilon_k, 0) = 0$ for only *finitely many* of those $k$. By proposition 3.12 ($k_0 = -1$),

$$(26) \qquad (\gamma^+ - \kappa_1(\gamma^+)) \nmid g_{\chi^\circ c}.$$

Since $g_\chi | h_\chi f_\chi$ and $h_{\chi^\circ c} | g_{\chi^\circ c}$, (24) implies that

$$(27) \qquad (\gamma^+ - \kappa_1(\gamma^+))|f_\chi.$$

The restriction of $\chi$ to $\Delta$ is $\omega$, hence $f_\omega$ is divisible by $\gamma^+ - \kappa_1(\gamma^+)$. Corollary 3.8 completes the proof of the main theorem.

# BIBLIOGRAPHY

Arthaud, N.

[Art] On Birch and Swinnerton-Dyer's conjecture for elliptic curves with complex multiplication I. Comp. Math. 37 (1978), 209-232.

Artin, E., Hasse, H.

[A-H] Die beiden Ergänzungssätze zum Reziprozitätsgesetz der $\ell^n$-ten Potenzreste im Körper der $\ell^n$-ten Einheitzwurzeln. Hamb. Abh. 6 (1928), 146-162.

Bashmakov, M.I.

[Ba] The cohomology of abelian varieties over a number field. Russian Math. Surveys 27 (1972), 25-70.

Bernardi, D., Goldstein, C., Stephens, N.

[B-G-S] Notes $p$-adiques sur les courbes elliptiques. J. Reine Angew. Math. 351 (1985), 129-170.

Birch, B., Swinnerton-Dyer, H.P.F.

[B-SD] Notes on elliptic curves. J. Reine Angew. Math. I, 212 (1963), 7-25; II, 218 (1965), 79-108.

Borel et al.

[Bo] Seminar on complex multiplication. Springer-Verlag LNM 21 (1966).

Brumer, A.

[Br] On the units of algebraic number fields. Mathematika 14 (1967), 121-124.

Cassou-Noguès, P.

[CN] $p$-adic $L$ functions for elliptic curves with complex multiplication I. Comp. Math. 42 (1981), 31-56.

Chandrasekharan

[Chand] Elliptic Functions. Springer-Verlag (1985).

Coates, J.

[C] Infinite descent on elliptic curves with complex multiplication. In: Arithmetic and Geometry, Birkhäuser, Progress in Math. 35 (1983), 107-137.

Coates, J., Goldstein, C.

[C-Go] Some remarks on the main conjecture for elliptic curves with complex multiplication. Am. J. Math. 103 (1983), 411-435.

Coates, J., Wiles, A.

[C-W1] On the conjecture of Birch and Swinnerton-Dyer. Inv. Math. 39 (1977), 223-251.

[C-W2] On $p$-adic $L$ functions and elliptic units. J. Austral. Math. Soc. (series A) 26 (1978), 1-25.

[C-W3] Kummer's criterion for Hurwitz numbers. In: Algebraic Number Theory, Kyoto 1976. Japan Society for the Promotion of Science (1977), 9-23.

Coleman, R.

[Col1] Division values in local fields. Inv. Math. 53 (1979), 91-116.

[Col2] The arithmetic of Lubin Tate division towers. Duke Math. J. 48 (1981), 449-466.

Cuoco, A.

[Cu] The growth of Iwasawa invariants in a family. Comp. Math. 41 (1980), 415-437.

Damerell, R.M.

[Da] $L$ functions of elliptic curves with complex multiplication. Acta Arith. I, 17 (1970), 287-301; II, 19 (1971), 311-317.

Deligne, P., Ribet, K.

[D-R] Values of abelian $L$-functions at negative integers over totally real fields. Inv. Math. 59 (1980), 227-286.

de Shalit, E.

[dS1] Relative Lubin Tate groups. Proc. of the A.M.S. 95 (1985), 1-4.

[dS2] The explicit reciprocity law in local class field theory. Duke Math. J. 53 (1986), 163-176.

[dS3] On monomial relations between $p$-adic periods. To appear in J. Reine Angew. Math.

Deuring, M.

[Deu] Die Klassenkörper der komplexen Multiplikation. Enzyclopädie Math. Wiss. Neue Aufl. Band I-2, Heft 10-II, Stuttgart (1958).

Gillard, R.

[Gi1] Unités elliptiques et fonctions $L$ $p$-adiques. Comp. Math. 42 (1981), 57-88.

[Gi2] Fonctions $L$ $p$-adiques des corps quadratiques imaginaires et de leurs extensions abéliennes. J. Reine Angew. Math. 358 (1986), 76-91.

Gillard, R., Robert, G.

[Gi-R] Groupes d'unités elliptiques. Bull. Soc. Math. France 107 (1979), 305-317.

Goldstein, C., Schappacher, N.

[Go-Sch] Séries d'Eisenstein et fonctions $L$ des courbes elliptiques à multiplication complexe. J. Reine Angew. Math. 327 (1981), 184-218.

Greenberg, R.

[Gre1] On the structure of certain Galois groups. Inv. Math. 47 (1978), 85-99.

[Gre2] On the Birch and Swinnerton-Dyer conjecture. Inv. Math. 72 (1983), 241-265.

[Gre3] On the critical values of Hecke $L$ functions for imaginary quadratic fields. Inv. Math. 79 (1985), 79-94.

Gross, B.

[Gr] Arithmetic of elliptic curves with complex multiplication. Springer-Verlag LNM 776 (1980).

Gross, B., Zagier, D.

[G-Z] Heegner points and derivatives of $L$-series. Inv. Math. 84 (1986), 225-320.

Hazewinkel, M.

[Haz] Formal groups and applications. Academic Press, New York (1978).

Iwasawa, K.

[Iw] On $\mathbf{Z}_\ell$ extensions of algebraic number fields. Ann. of Math. (2) 98 (1973), 246-326.

[Iw2] Lectures on $p$-adic $L$ functions. Ann. of Math. Studies 74, Princeton (1972).

[Iw3] Local class field theory. Oxford University Press, Oxford (1986).

[Iw4] Explicit formulas for the norm residue symbol, J. Math. Soc. Japan 20 (1968), 151-164.

Katz, N.

[K1] $p$-adic interpolation of real analytic Eisenstein series. Ann. of Math. (2) 104 (1976), 459-571.

[K2] $p$-adic $L$ functions for CM fields. Inv. Math. 49 (1978), 199-297.

Kubert, D., Lang, S.

[K-La] Modular Units. Springer-Verlag (1981).

Kubota, T., Leopoldt, H.W.

[K-Le] Eine $p$-adische Theorie der Zetawerte. I. Einführung der $p$-adische Dirichletschen $L$-funktionen. J. Reine Angew. Math. 214/215 (1964), 328-339.

Kummer, E.

[Kum] Über die Erganzungssätze zer den allgemeinen Reciprocitätsgesetzen. J. Reine Angew. Math. 44 (1852), 93-146. Collected paper I, 485-538.

Lang, S.

[La] Elliptic Functions. Addison-Wesley (1973).

[La2] Algebraic Number Theory. Addison-Wesley (1970).

[La3] Cyclotomic Fields. Springer-Verlag GTM 59 (1978).

Lubin, J., Tate, J.

[L-T] Formal complex multiplication in local fields. Ann. of Math. 81 (1965), 380-387.

Manin, J.

[Man] Periods of cusp forms, and $p$-adic Hecke series. Mat. Sbornik (N.S.) 92 (1973), 378-401. English trans.: Math. USSR-Sb. 21 (1973), 371-393.

Manin, J., Višik, M.

[M-V] $p$-adic Hecke series of imaginary quadratic fields. Math. Sbornik (N.S.) 95 (1974), 357-383. English trans.: Math. USSR-Sb. 24 (1974), 345-371.

Mazur, B.
[Maz1] Analyse $p$-adique. Bourbaki (1972).
[Maz2] Rational points on abelian varieties with values in towers of number fields. Inv. Math. 18 (1972), 186-266.

Mazur, B., Swinnerton-Dyer, H.P.F.
[M-SD] Arithmetic of Weil curves. Inv. Math. 18 (1972), 183-266.

Mumford, D.
[Mum] Abelian Varieties. Oxford Univ. Press (1974).

Perrin-Riou, B.
[PR1] Arithmétique des courbes elliptiques et théorie d'Iwasawa. Bull. Soc. Math. France, Memoire (N.S.) 17 (1984).
[PR2] Points de Heegner et dérivées de fonctions $L$ $p$-adiques. C.R. Acad. Sc. Paris, t. 303, série I, (1986) 165-168.

Ramachandra, K.
[Ra] Some applications of Kronecker's limit formulas. Ann. Math. 80 (1964), 104-148.

Robert, G.
[R] Unités elliptiques. Bull. Soc. Math. France, Mémoire 36 (1973).

Rohrlich, D.
[Roh] On $L$ functions of elliptic curves and anti-cyclotomic towers. Inv. Math. 75 (1984), 383-408.

Rubin, K.
[Ru1] Elliptic curves with complex multiplication and the conjecture of Birch and Swinnerton-Dyer. Inv. Math. 64 (1981), 455-470.
[Ru2] Descents on elliptic curves with complex multiplications. MSRI preprint (1986).
[Ru3] In preparation (1986).

Serre, J.P.
[Se] Local class field theory. In: Algebraic Number Theory, Cassels, J.W.S. and Fröhlich, A. eds., Academic Press, London (1967), 128-161.
[Se2] Sur le résidu de la fonction zêta $p$-adique d'un corps de nombres. C.R. Acad. Sci. Paris, Sér. A. 287 (1978), 183-188.
[Se3] Corps Locaux. Hermann, Paris (1962).

Serre, J.P., Tate, J.
[Se-Ta] Good reduction of abelian varieties. Ann. Math. 88 (1968), 492-517.

Shimura, G.
[Sh] Introduction to the arithmetic theory of automorphic functions. Princeton Univ. Press (1971).
[Sh2] The special values of the zeta functions associated with cusp forms. Comm. Pure App. Math. 29 (1976), 783-804.

Siegel, C.L.
[Sie] Lectures on advanced analytic number theory. Tata Inst. of Fund. Research (1961).

Silverman, J.
[Sil] The arithmetic of elliptic curves. Springer-Verlag GTM 106 (1986).

Stephens, N.M.
[St] The diophantine equation $x^3 + y^3 = Dz^3$ and the conjecture of Birch and Swinnerton-Dyer. J. Reine Angew. Math. 231 (1968), 121-162.

Tate, J.
[Ta] The arithmetic of elliptic curves. Inv. Math. 23 (1974), 179-206.
[Ta2] $p$-divisible groups. In: Proceedings of a conference on Local Fields, T.A. Springer, ed. Springer-Verlag (1967).
[Ta3] Les conjectures de Stark sur les fonctions $L$ d'Artin en $s = 0$. Birkhäuser, Progress in Math. 47 (1984).
[Ta4] On the conjectures of Birch and Swinnerton-Dyer and a geometric analog. Sém. Bourbaki 306 (1966).

Tilouine, J.
[Ti] Fonctions $L$ $p$-adiques à deux variables et $\mathbf{Z}_p^2$-extensions. Bull. Soc. Math. France, 114 (1986), 3-66.

Washington, L.
[Wa] Introduction to cyclotomic fields. Springer-Verlag GTM 83 (1982).

Weber, H.
[Web] Lehrbuch der Algebra III. Strassburg (1908).

Weil, A.
[We] On a certain type of characters of the idèle-class group of an algebraic number field. In: Proc. Int. Symp. on Alg. Number Th. Tokyo (1955) p. 1-7.
[We2] Elliptic functions according to Eisenstein and Kronecker. Springer-Verlag (1976).
[We3] Dirichlet series and automorphic forms. Springer-Verlag LNM 189 (1971).

Whittaker, E.T., Watson, G.N.
[W-W] A Course on Modern Analysis (4th edition). Cambridge Univ. Press, Cambridge (1958).

Wiles, A.
[Wi] Higher explicit reciprocity laws. Ann. Math. (2) 107 (1978), 235-254.

Wintenberger, J-P.
[Win] Structure galoisienne de limites projectives d'unités locales. Comp. Math. 42 (1981), 89-103.

Yager, R.

[Ya1] On two variable $p$-adic $L$ functions. Ann. Math. 115 (1982), 411-449.

[Ya2] $p$-adic measures on Galois groups. Inv. Math. 76 (1984), 331-343.

[Ya3] A Kummer criterion for imaginary quadratic fields. Comp. Math. 47 (1982), 31-42.

Haran, S.

[Har] p-adic L functions for modular forms. Preprint, Max Planck Institut, Bonn (1986). To appear in Comp.Math.

## Index of Notation

Page numbers refer to first occurence(s). If the meaning of a symbol slightly varies, this should be clearly indicated in the text.

### 0. General.

| | |
|---|---|
| $R^x$ | units of $R$ |
| $R[[T]]$ | power series ring over $R$ |
| $\bar{k}$ | algebraic closure of $k$ |
| $\mathbf{Z}_p, \mathbf{Q}_p$ | $p$-adic integers and $p$-adic numbers |
| $\mathbf{C}_p$ | completion of $\overline{\mathbf{Q}}_p$ |
| $F_P$ | completion of $F$ at $P$ |
| $N_{F/K}, Tr_{F/K}$ | norm, trace |
| $\mu_m$ | $m^{th}$ roots of unity |
| $\hat{G}$ | character group of $G$ |
| $\mathbf{N}$ | absolute norm |
| $M[n]$ | $Ker(n: M \to M)$ |
| $M^G, M_G$ | $G$-invariants, $G$-coinvariants |
| $M^\chi, M_\chi$ | $\chi$-eigenspace, $\chi$-coeigenspace |
| $\prod'$ | the prime means that an obvious exception should be made to the index set |
| $rk\ M$ | rank of a finitely generated abelian group |
| $E(F)$ | $F$-valued points of $E$ |
| $\bar{a}$ | complex conjugate of $a$ |

### 1. Fields, rings and ideals.

**6. Formal groups and related symbols (especially ch. I).**



**7. Elliptic curves and special functions.**

# Perspectives in Mathematics

Vol. 1 J.S. Milne, *Arithmetic Duality Theorems*

Vol. 2 A. Borel et al., *Algebraic D-Modules*

Vol. 3 Ehud de Shalit, *Iwasawa Theory of Elliptic Curves with Complex Multiplication*